Ethics of Science and Technology Assessment

Founding Editor

Carl Friedrich Gethmann

Volume 54

The series *Ethics of Science and Technology Assessment* focuses on the impact that scientific and technological advances have on individuals, their social lives, and on the natural environment. Its goal is to cover the field of Science and Technologies Studies (STS), without being limited to it. The series welcomes scientific and philosophical reviews on questions, consequences and challenges entailed by the nature and practices of science and technology, as well as original essays on the impact and role of scientific advances, technological research and research ethics. Volumes published in the series include monographs and edited books based on the results of interdisciplinary research projects. Books that are devoted to supporting education at the graduate and post-graduate levels are especially welcome.

Pablo López-Silva
Editor

Contextualizing Neuroprotection: Latin American Perspectives on the Impact of Neurotechnological Development in Life and Society

Springer

Editor
Pablo López-Silva
School of Psychology
University of Valparaíso
Valparaíso, Chile

Institute of Complex Systems
Valparaíso, Chile

Millennium Institute for Research
in Depression and Personality, MIDAP
Santiago, Chile

Research Centre in Cognition
and Language Development
Universidad de Valparaíso
Valparaíso, Chile

The Collaborating Centre for Values-based
Practice in Health and Social Care
St Catherine's College
University of Oxford
Oxford, UK

ISSN 1860-4803 ISSN 1860-4811 (electronic)
Ethics of Science and Technology Assessment
ISBN 978-3-031-96054-3 ISBN 978-3-031-96055-0 (eBook)
https://doi.org/10.1007/978-3-031-96055-0

This Springer imprint is published by the registered company Springer Nature Switzerland AG
The registered company address is: Gewerbestrasse 11, 6330 Cham, Switzerland

Acknowledgments

In October 2019, I received an interesting phone call. Before that, I had no idea what a neuroright was. Of course, I was aware of the many ethical concerns raised by the rapid development of neurotechnological applications beyond the medical field. My love for science fiction and fantasy movies had always fueled my curiosity about the relationship between humans and machines, as well as my endless reflections on what it means to be human. A dear colleague of mine had been contacted to recommend someone willing to comment on the keynote talk of a workshop titled *Is It Time for Neurorights?* For some reason, he recommended me—and to this day, I'm still not sure why. The idea was to share some thoughts with an interdisciplinary group of researchers on Rafael Yuste's proposal within the broader discussion of how to protect people from the potential misuse of neurotechnologies. Since that day, a lot has happened, and the discussion has thrived on many fronts, bringing together researchers from all over the world to exchange ideas on a matter that directly affects citizens across different countries. Papers have been published, compilations have been edited, projects have been funded, and legal changes have been debated. I am deeply grateful to have had the opportunity to be part of this process.

Over the past few years, I've met amazing academics, but more importantly, good people. I am especially grateful to Abel Wajnerman and Raúl Madrid for the time they have spent discussing ideas with me, for their friendship, and for their academic rigor. I also thank Paulina Ramos, Luca Valera, Adrian Palacios, and many others who have supported and encouraged me, allowing me to develop and share my ideas in this debate. To all the contributors of this compilation: thank you for the work you have done in your respective countries and for the ideas you have fostered. I also thank the Universidad de Valparaiso School of Psychology for providing a supportive space to develop and refine my ideas and my student Pablo Alvarez for his dedication to academic tasks with remarkable sense of responsibility. Finally, I shall thank my wife—thank you for your unwavering support throughout my academic career. *Contextualizing Neuroprotection* was born from a genuine desire to share the struggles and insights we have encountered in addressing the many worries raised by the use of neurotechnology in current society. It is my hope that the work here presented can inform and foster further reflections on concerns that may shape our near future.

Pablo López-Silva

Contents

Editor and Contributors

About the Editor

Pablo López-Silva is Professor of psychology at the Universidad de Valparaiso Faculty of Social Sciences (School of Psychology) and Research Professor at the Institute of Philosophy at the same institution. After earning undergraduate and professional degrees in psychology in Chile, he completed an MRes and a Ph.D. in philosophy at the University of Manchester, UK. Currently, he is also Young Research Fellow at the Millennium Institute for Research in Depression and Personality (MIDAP-Chile), Research Fellow at the Valparaiso Institute for Complex Systems, Associate Researcher at the Universidad de Valparaiso Research Centre for Cognition and Language Development, and Project Partner at the Collaborating Centre for Values-Based Practice in Health and Social Care at St Catherine's College, University of Oxford, UK.

Contributors

Florencia Álamos Centro de Bioética UC, Facultad de Medicina, Pontificia Universidad Católica de Chile, Santiago, Chile

Gustavo A. Arocena Law School, Universidad Nacional de Córdoba, Córdoba, Argentina

Gabriela Arriagada-Bruneau Instituto de Éticas Aplicadas (IEA), Instituto de Ingeniería Matemática y Computacional (IMC), Pontificia Universidad Católica de Chile, Santiago, Chile;
Centro Nacional de Inteligencia Artificial (CENIA), Santiago, Chile

Diego Borbón Center for Studies on Genetics and Law, Research Group on Biological Sciences and Law, Universidad Externado de Colombia, Bogotá, Colombia;

Institute of Biomedical Ethics and History of Medicine (IBME), University of Zurich, Zurich, Switzerland

Roberto Cippitani Chair 'ISAAC' (Individual Rights in Scientific Research) at Universidad Nacional de Educación a Dinstancia, Madrid, Spain;
INDEPAC - Instituto Nacional de Estudios Superiores en Derecho Penal, Ciudad de México, México;
Consiglio Nazionale Delle Ricerche, CNR-IFAC, Firenze, Italy

Isabel Cornejo-Plaza Neurometa; Research Group on Neuro-Rights, Artificial Intelligence and Metaverse; DDLAB (Innovation Laboratory, Technologies and Frontiers of Legal Sciences), Universidad Autónoma de Chile, Santiago, Chile

Eduardo Demetrio Crespo Universidad de Castilla-La Mancha, Toledo, Spain

Faustino García de la Torre García Universidad de Castilla-La Mancha, Toledo, Spain

Jorge Ignacio Fuentes Centro de Bioética UC, Facultad de Medicina, Pontificia Universidad Católica de Chile, Santiago, Chile

Pablo López-Silva Escuela de Psicología, Universidad de Valparaíso, Valparaíso, Chile;
Instituto de Sistemas Complejos de Valparaíso, Valparaíso, Chile;
Centro de Investigación del Desarrollo en Cognición y Lenguaje, Universidad de Valparaíso, Valparaíso, Chile;
The Collaborating Centre for Values-Based Practice in Health and Social Care, St Catherine's College, Universidad de Oxford, Oxford, UK;
School of Psychology and Research Centre in Cognition and Language Development, Universidad de Valparaíso, Valparaíso, Chile;
Millenium Institute for Research in Depression and Personality (MIDAP-Chile), Valparaíso Institute for Complex Systems, Valparaíso, Chile

Raúl Madrid Facultad de Derecho. Facultad de Filosofía, Law, Science & Tecnology Program, Pontificia Universidad Católica de Chile, Santiago, Chile;
Faculty of Law and Faculty of Philosophy, Law, Science and Technology Program, Pontificia Universidad Católica de Chile, Santiago, Chile

José Angel Marinaro Universidad Nacional de La Matanza, San Justo, Argentina

Enrique Cáceres Nieto National Autonomous University of México Legal Research Institute, Maestro Mario de La Cueva S/N Ciudad Universitaria, Ciudad de México, México

Carmen Patricia López Olvera National Autonomous University of México Legal Research Institute, Maestro Mario de La Cueva S/N Ciudad Universitaria, Ciudad de México, México

Camila Pintarelli State of São Paulo's Attorney General's Office, Sao Paulo, Brazil

Paulina Ramos Vergara Centro de Bioética UC, Facultad de Medicina, Pontificia Universidad Católica de Chile, Santiago, Chile

Gabrielle Bezerra Sales Sarlet Pontifícia Universidade Católica do Rio Grande do Sul, Porto Alegre, Brasil

Nicolás Alejandro Serrano Universidad de Buenos Aires, Buenos Aires, Argentina

Abel Wajnerman Paz Instituto de Éticas Aplicadas (IEA), Pontificia Universidad Católica de Chile, Santiago, Chile

Abel Wajnermann Instituto de Éticas Aplicadas (IEA), Pontificia Universidad Católica de Chile, Santiago, Chile

Lucas Reckziegel Weschenfelder Pontifícia Universidade Católica do Rio Grande do Sul, Porto Alegre, Brasil

Danielle Zaror-Miralles Santiago, Chile

List of Figures

Towards a Contextualized Examination of the Impact of Neurotechnological Development in Life and Society: Conceptual and Practical Lessons

Pablo López-Silva

Abstract This chapter introduces and contextualizes some of the main debates regarding the impact of neurotechnological developments in Latin America over the last years. In its first part, the chapter makes sense of core conceptual lessons that local researchers have learnt from their active involvement in the discussion of these issues in their local governments and academic communities. In its second part, the chapter focuses on a number of practical cases derived from the Latin American experience on discussions about neuroprotection, neurorights, neurolaw regulations, and artificial intelligence policy.

Keywords Neuroprotection · Neurorights · Latin America · Neurotechnology

Neurotechnology is an umbrella term typically used to refer to a broad and heterogenous spectrum of methods, systems and instruments able to establish direct connections with the human brain through which neuronal activity can be recorded and/or influenced (Ienca, 2021). Over the last years, collaborative international research initiatives have led to the development of novel methods to record and manipulate neural activity with unprecedented specificity (Koroshetz et al., 2018; Ngai, 2024). Considering its nature, the rapid development of neurotechnological applications has motivated a number of practical, conceptual, and ethical worries all around the world over the years. Initially, most of these worries emerged in the context

P. López-Silva (✉)
School of Psychology, Universidad de Valparaíso, Valparaiso, Chile
e-mail: pablo.lopez.silva@gmail.com

Millenium Institute for Research in Depression and Personality (MIDAP-Chile), Santiago, Chile

Universidad de Valparaíso Research Centre in Cognition and Language Development, Valparaíso, Chile

Valparaíso Institute for Complex Systems, Valparaíso, Chile

Collaborating Centre for Value-Based Research, St Catherine's College, University of Oxford, Oxford, UK

P. López-Silva (ed.), *Contextualizing Neuroprotection: Latin American Perspectives on the Impact of Neurotechnological Development in Life and Society*, Ethics of Science and Technology Assessment 54, https://doi.org/10.1007/978-3-031-96055-0_1

of medical research in debates about how neurotechnologies could impact clinical practice, agency, autonomy, consent, and many other issues (see, Yuste et al., 2017; Yuste, 2019; Ienca, 2021; Álamos et al., 2022). However, the growing production of other types of neurotechnological applications with direct and indirect access to neural data (commercial, military, managerial, etc.) has forced governments to take a stance on how the law should protect citizens from the potential misuses of such technologies.

Already in 2013, former US President Barack Obama called attention upon the potential impact of neurotechnologies in society, emphasizing the need to address questions about: "privacy, personal agency, and moral responsibility for one's actions," along with important issues regarding "stigmatization and discrimination based on neurological measures of intelligence or other traits; and questions about the appropriate use of neuroscience in the criminal-justice system" (Presidential Commission for the Study of Bioethical Issues, 2014). In response to these concerns, in 2017, Columbia's MorningSide Group introduced the idea that advancements in neurotechnology—both medical and commercial—should be accompanied by the development of legal safeguards to protect 'people's privacy, identity, agency, and equality' (Yuste et al., 2017, 159). The underlying premise was that existing legal frameworks were not well-equipped to address the novel threats neurotechnology poses to neural data privacy and other fundamental values. These safeguards, broadly referred to in the literature as *neurorights*, are intended to protect individuals from the potential risks associated with neurotechnological developments (Zuñiga-Fajuri et al., 2021, 166). Since then, the discussion about how to protect citizens from the misuses of neurotechnologies has expanded significantly in the world.

The impact of the mass use of neurotechnological applications with direct and indirect access to neural data could be multilevel and global (López-Silva & Valera, 2022a, 2022b). From behavioural and psychological changes in individuals, to the reproduction of social, economic, and epistemic asymmetries, the irruption of neurotechnologies in everyday life calls for a conceptual, legal, and practical analyses to inform policymaking and ongoing changes in legal frameworks worldwide. It is in this context that Latin America has been a privileged witness of the type of complexities that the discussion of these issues involves (López-Silva & Madrid, 2022, 2024; Borbon, 2025; Herrera-Ferrá et al., 2025). For the right or wrong reasons, with the time Latin America became a protagonist when it comes to the discussion of specific legal issues surrounding the analysis of the impact of neurotechnologies in society (García-López et al., 2021; López-Silva & Valera, 2022a, 2022b; Muñoz & Marinaro, 2023; López-Silva et al., 2024). Along with the development of these conceptual debates—but not always in synchrony -, Latin America has been undertaking some practical—yet problematic—actions, such as a constitutional change in Chile, and a new law in Brazil (parallel to the Spanish project) both aiming at protecting neural data at different levels and dimensions from the potential misuses of neurotechnologies (for critical examinations, *see* López-Silva & Madrid, 2021; Possa & Schertel-Mendes, 2025). The fact that Chile was the first country in the world that debated a neuroprotective law has encouraged local researchers from Colombia, México, Argentina, Uruguay, Brazil, and other countries in the region to build a strong collaborative network focused on discussing how local experiences could positively or negatively inform the development of the many debates occurring in the area in different countries.[1] In 2022, as part of Springer's *Ethics of Science and Technology Assessment* series, Pablo López-Silva and Luca Valera published the first compilation in the world entirely dedicated to examining core conceptual and practical problems emerging from the use of neurotechnologies in Latin American society, and the world. Over the years, the Latin American experience in the discussion and development of neuroprotective laws and neurorights has both enlightened—and, in some cases, obscured—both European and North American forums, creating important opportunities worldwide to debate and collaborate on matters of common concern.

Being a pioneer in a field presents both advantages and challenges. While it allows for innovation and the establishment of novel paradigms and ideas, it also entails navigating uncharted territory, facing resistance, and lacking established support conceptual frameworks. Over the last 3 years, a number of Latin American researchers have participated in discussions all around the world informing not only through their academic expertise in fields such as medicine, philosophy, law, and psychiatry, but also, from having the experience of being part of the discussion of such changes in law and ethical guidelines in their local governments.[2] Some of these researchers have even become global referents when it comes to the discussions about how to regulate the law in order to protect people from the uses and misuses of neurotechnological developments. However, the road has not been easy, and a number of harsh lessons have been learnt in the making of these discussions. Additionally, due to a number of socio and geo economic conditions, epistemic asymmetries, and connectivity issues, sometimes, the contributions of Latin American researchers have been marginalized in global academic discussions, particularly when such debates are primarily situated in other continents

To build on the work of López-Silva and Valera (2022a, 2022b), the compilation *'Contextualizing Neuroprotection: Latin American Perspectives on the Impact of Neurotechnological Development in Life and Society'* brings forward the informed-by-experience insights of Latin America-based researchers on the debate about how to protect citizens from the risks raised by neurotechnological applications. It gathers conceptual and practical lessons we have learnt in the making of a complex and obscure process, namely, the discussion of the impact of neurotechnologies in society and the debate about how legal frameworks should—or should not—be changed in order to deal with these problems. The contributions contained in this collection offer an updated set of conceptual and practical lessons that have the potential of informing the development of current global discussion on the philosophy, law, and ethics of neuroprotection from the perspective of these researchers' own intellectual paths and

[1] The creation of the *Ibero-American Network in Neuroright and Neuroethics* hosted by the Universidad Externado de Colombia is the product of this collaborative stance towards issues of mutual interest in the continent.

[2] Most authors of the chapter contained in this compilation have been part of the discussions on neurorights, neurolaw, and neuroprotection in their respective countries.

personal experience. With this, the compilation put its attention on how the different collaborations offered by Latin America-based researchers could offer practical, conceptual, and legal inputs that might inform decision-making in communities all around the world that are dealing with the same type of worries and ethical challenges.

To provide a comprehensive and cohesive structure, this compilation is organized into two main parts. The first part examines the conceptual insights gained by Ibero-American researchers in discussions on the impact of neurotechnology on issues such as free will, criminal law, the concept of personality, and, importantly, neurorights. In Chapter "Free Will Skepticism, Neurorights, and Coercive Moral Neuroenhancements: An Approach from Neuroabolitionism", Diego Borbón examines the debate occurring at the intersection of free will skepticism, neurorights, and the ethics of coercive moral neuroenhancements. Briefly, hard incompatibilism questions whether individuals possess the control necessary for moral responsibility. If free will is denied, some may argue for coercive neuroenhancements to curb antisocial behavior. This raises concerns about neurorights, which seek to protect values such as cognitive liberty and consent. Borbón argues that, in this context, current neurorights frameworks are insufficient and could undermine human rights. A hard incompatibilist perspective should reassess moral enhancements and neurorights, advocating for more humane approaches aligned with penal neuroabolitionism. This view calls for the abolition of criminal law and punishment, emphasizing respect for human dignity, cognitive freedom, bodily integrity, autonomy, and consent. In Chapter "Neurotechnologies, Human Rights and Criminal Law", Eduardo Demetrio Crespo and Faustino García de la Torre García examine the impact that recent discoveries in neuroscience have on the basic assumptions of criminal liability as it deals with the subjective aspects of human behaviour. They claim that neurotechnologies, which may prove to be of utility to the justice system, could also serve to commit crimes. This contribution sets out the main challenges for Criminal Law systems in the wake of advances in the neurosciences and the development of technologies applied to the study of the human brain. In the final section, the authors provide an examination of some relevant discussions about the epistemological and axiological status of criminal law in the era of neurotechnologies. Keeping a conceptual focus, in Chapter "Personal identity and personality neurorights: reconceptualize to protect", José Ángel Marinaro examines the links between the legal concepts of identity and personality. In the literature, both concepts are often treated collectively, separately, or indistinctly, given that their neurotechnological underpinning are considered identical. To the author, clarifying the link between these concepts will be the basis of a schematization where personality in its semantic field has the aptitude to incorporate the notion of (personal) identity. Marinaro claims that this allows us to classify and systematically place psychological continuity and its close relative, identity self-perception, two core notions when it comes to the discussion of neuroprotective laws and regulation. In Chapter "Deconstructing the Mind: Neurotechnology, Mental Privacy, and the Case for Neurorights", Raúl Madrid introduces a more specific discussion that will be directly linked with the themes of part 2. Madrid explores the challenges posed by neurotechnology and artificial intelligence to traditional understandings of mental privacy and human rights from

Derrida's deconstructive approach. By situation some of the main worries about the use of neurotechnology Derrida's critique of metaphysical oppositions (specifically the inside/outside dichotomy and his non-binary concept of the "trace"), the chapter suggests that neurotechnology deconstructs traditional notions of the mind as an inviolable interior space. Madrid argues that this transformation demands urgent legal responses, such as the incorporation of "neurorights" into constitutional and statutory frameworks to protect mental integrity. With this, the author underscores the importance of a robust legal infrastructure to address risks including neuroprivacy violations, manipulation of free will, and neurodiscrimination, as well as the broader philosophical implications of rethinking the human mind as a material system open to external interpretation and influence. The final chapter of Part 1 of the compilation introduces the reader to a contextualized perspective on the problem of predicting criminal behavior considering the case of Argentine penitentiary law. To do so, in Chapter "Criminal Law, Prognosis of Criminal Behavior, Neuroscience and the Case of Psychopathology", Gustavo Arocena describes clinical and statistical methods used by criminal courts and examines their virtues and disadvantages. After this, the author postulates the convenience of resorting to the use of composite prediction protocols. Finally, the author focuses on the possible contributions of neuroscientific tests in the prediction of violent, antisocial, or criminal behavior

Shifting to a more practical perspective, Part 2 of the compilation gathers a series of specific case studies from where the reader could learn important lessons when it comes to discussions about the relation between neurotechnologies and health systems, digital law, criminal behavior neurolaw, consent, artificial intelligence, and, of course, the neurorights debate in the region. In its final part, this part of the book offers comparative analyses for a more comprehensive understanding of the way in which the neurorights debate has developed in Latin America compared to the case of the debate about how to regulate artificial intelligence. In Chapter "Neurotechnologies and Inequality: Insights From Latin America", Nicolás Serrano introduces a more specific treatment of the concept of neurorights and claims that the peripherical socio-economic position of most Latin American countries give local researchers a particular perspective on the potential consequences of emerging neurotechnologies in society. The author argues that this perspective can bolster international discourse by highlighting the importance of specific topics meaningfully informing international analysis with new proposals based on regional insights. Following this spirit, the following six chapters dive into the way in which local experience in Latin America have taught us important practical and theoretical lessons regarding regulation of neurotechnologies and their impact in different aspects of society. First, Chapter "Advances Towards a Legal Theory of Neurolaw Based on the Case of Mexico" offers an updated examination of the state of the art of the discussion about neurorights in Mexico. After analyzing, the conceptual scaffolding of the notion of 'neurolaw' and 'neurorights', Enrique Cáceres Nieto and Carmen Patricia López Olvera explores three main legal positions on the regulation of neurorights, namely, an exclusive one that does not call for specific categories since they are protected by existing positive law, an inclusive position that underscores the need for new regulatory frameworks to deal with the threat of potential neurotechnological misuses; and a hermeneutic position that adapts existing law to these new conditions through interpretation. After this, the authors examine neurolaw and neurorights-related activities in Mexico to propose a general theory of law with the potential of guiding academic discourse, legislation and case law work on neurorights both nationally and internationally. Moving further south, the following two chapters offer two case studies of the way in which Brazil has faced the discussion about the impact of neurotechnologies in society and neurorights. In Chapter "The Right to Health in Brazil and the Prospects for Using Neurotechnology", Gabrielle Bezerra Sales Sarlet and Lucas Reckziegel Weschenfelder explore the evolving legal framework governing neurotechnology within Brazil's digitalized health system. The authors provide a contextualized overview of state policies and regulatory frameworks related to telemedicine and human research, highlighting their potential adaptations to accommodate neurotechnologies. In its final sections, the authors address the emerging perspective on neurorights, with a focus on Brazilian and broader Latin American legal initiatives and conclude by underscoring the imperative of a cross-updating of the regulatory system, particularly concerning healthcare, medical practice, and research, with specific attention to neurotechnologies incorporating artificial intelligence. In Chapter "Neuroscience, Neuroprotection, and Neurorights: Human Brain Rights and the Brazilian Constitutional Advances", Camilla Pintarelli explores some of the theoretical foundations for the idea that scientific and technological development must serve humanity and respect life, equality and freedom. After examining the social evolutionary milestones related to this subject and the ethical impulses derived from them, the author examines the current legal stage of the protection of the human mind in the international context. Finally, the author analyzes the possibilities that Brazil has regarding the legal treatment of these issues, using a descriptive exposition and attentive to the legislative changes underway in the country.

The final section of Part 2 focuses on the lessons that researchers based in Chile have learnt from the many complexities of being active participants in the discussion of the first constitutional amendment for the protection of the human mind, and the first bill for neurorights in the world. After the initial enthusiasm generated by these legal initiatives, local researchers found a number of conceptual and practical difficulties. Mistakes were made, and dissention among academics and politicians led to intense debates, some of them still unresolved. Since then, a number of lessons have been learnt, and this section gathers some of the main ones. In Chapter "Neural Data and Neuroprotection in the Chilean Constitution Reform Bill: Critical Considerations", Pablo López-Silva and Raul Madrid examine in detail the core message proposed by the constitutional reform that explicitly integrates legal safeguards against the misuses of medical and commercial neurotechnological applications in Chile (Law N° 21.383). After offering some critical considerations about the content of the bill, the authors examine the way in which it is aimed to incorporate the so-called 'neurorights' to the Chilean legal order. In the same vein, in Chapter "A Situated Approach to Neurorights Legislation in Chile" Jorge Fuentes, Paulina Ramos Vergara, Abel Wajnerman, Pablo López-Silva, and Florencia Álamos, argue that we should avoid introducing new fundamental rights in any legal framework whenever possible, focusing on the case of neurorights in Chile. The authors call this the situated approach to neurorights because the issue about protecting the mind without introducing new rights is a question that should be asked in each legal framework separately. Considering the Chilean case, the authors provide a critical assessment of local regulatory proposals and show that, as a result of the local legislative debate, they have evolved from a universalist stance towards a much more situated view. In Chapter "Understanding Consent: Challenges in the Age of Neurotechnologies", Danielle Zaror-Miralles analyses some of the weaknesses in the debate about how to regulate emerging technologies. The author argues that, in addition to be a redundant regulation—because the risks it is supposed to protect are covered by other fundamental rights –, neurorights in Chile have been designed with a defective regulatory technique that builds all their strength on consent. The problem, Zaror-Miralles argues, is that this enabling circumstance can only be delivered by healthy people. As a consequence, people with some level of deterioration in mental or brain activity—who benefit most from neurotechnologies—are in practice not only left out of the progress of science but will also experience lower levels of access to the benefits guaranteed by virtue of the right of access to health. Finally, Chapters "Understanding Consent: Challenges in the Age of Neurotechnologies" and "Principles of Research with Neurotechnologies and Artificial Intelligence: A Comparative Analysis of Latin American and European Laws" offer fruitful comparative analyses between the current development of the discussion of the ethics of artificial intelligence and the way in which the neurorights debate has evolved over the years. Isabel Cornejo-Plaza and Roberto Cippitani examine the ethical and legal implications of neurotechnologies and AI, particularly in research activities where such technologies are increasingly being used. They focus on the lack of explicit legislation on neurotechnology use in Europe, and how this contrast with but other legal initiatives regarding artificial intelligencc, such as the Artificial Intelligence Act and the General Data Protection Regulation. They authors observe that both Latin American and EU legal systems do not tend to focus their attention on the use of neurotechnologies and AI in research activities and offer some critical considerations regarding this issue. Finally, in Chapter "Principles of Research with Neurotechnologies and Artificial Intelligence: A Comparative Analysis of Latin American and European Laws", Gabriela Arriagada-Bruneau, Abel Wajnermann and Pablo López-Silva explore a potential way of making sense of the socio-political dynamics of the neurorights debate by looking into the evolution of the debate about artificial intelligence regulation in Chile. A study led by Humeres et al. (2025) identifies three key controversies in Chile's AI regulation process, namely, deflection of technology liability, instrumentalization of policy, and moralization of technology use. The authors explore how this analysis focused on sources of controversy could inform the neurorights debate currently in need of a systematic socio-political analysis. They argue that the application of this framework could help categorize the diverse positions within the discussion, providing a more comprehensive understanding of its development.

Considered as whole, the compilation *'Contextualizing Neuroprotection: Latin American Perspectives on the Impact of Neurotechnological Development in Life and Society'* collects a set of practical and conceptual lessons that researchers actively involved in debates on neuroprotection, neurorights, and artificial intelligence regulation in Latin America have learnt from experience for the last 3 years. Rather than proposing a clear path to be followed, the compilation is meant to motivate cautious consideration of ethical issues that impact our present and future life as individuals and as society. It is my hope that the work and time that researchers from México, Colombia, Brazil, Argentina, Chile, and Spain have dedicated to make this compilation possible further inspire and stimulate collaborative interdisciplinary work taking the Latin American experience as an example of how to do, and, many times, how not to do certain things.

References

Álamos, M. F., et al. (2022). A technocratic oath. In López-Silva, P., & Valera, L. (Eds.), *Protecting the mind. Ethics of science and technology assessment* (Vol. 49). Springer. https://doi.org/10.1007/978-3-030-94032-4_14

Borbón, D. (2025). What a neurorights legislation should not look like: The case of the Latin American parliament. *Frontiers in Neuroscience, 18*, 1514338. https://doi.org/10.3389/fnins.2024.1514338

García-López, E., Muñoz, J. M., & Andorno, R. (2021). Editorial: Neurorights and mental freedom: Emerging challenges to debates on human dignity and neurotechnologies. *Frontiers in Human Neuroscience, 15*, 823570. https://doi.org/10.3389/fnhum.2021.823570

Herrera-Ferrá, K., Muñoz, J. M., Becerril, A., García-López, E., Marinaro, J. Á., Sánchez Hernández, L. R., Alonso Olamendi, J. A., & Soto Ruiz, A. L. (2025). The regulation of neurotechnology: The neurorights bill in Mexico. *The Lancet. Psychiatry, 12*(2), 88–90. https://doi.org/10.1016/S2215-0366(24)00286-4

Ienca, M. (2021). On neurorights. *Frontiers in Human Neuroscience, 15*, 701258.

Koroshetz, W., Gordon, J., Adams, A., Beckel-Mitchener, A., Churchill, J., Farber, G., Freund, M., Gnadt, J., Hsu, N. S., Langhals, N., Lisanby, S., Liu, G., Peng, G. C. Y., Ramos, K., Steinmetz, M., Talley, E., & White, S. (2018). The state of the NIH BRAIN initiative. *The Journal of Neuroscience, 38*(29), 6427–6438. https://doi.org/10.1523/JNEUROSCI.3174-17.2018

López Silva, P., & Madrid, R. (2022). Acerca de la protección constitucional de los neuroderechos : La innovación chilena [en línea]. *Prudentia Iuris., 2022*(94), 39–68. https://doi.org/10.46553/prudentia.94.2022

López-Silva, P., & Madrid, R. (2021). Sobre la Conveniencia de incluir los Neuroderechos en la Constitución o en la Ley. *Revista Chilena De Derecho y Tecnología, 10*(1), 49–72. https://doi.org/10.5354/0719-2584.2021.56317

López-Silva, P., Wajnerman-Paz, A., & Molnar-Gabor, F. (2024). Neurotechnologies and the Problem about How to Protect Mental Privacy. *Neuroethics, 17*, 31. https://doi.org/10.1007/s12152-024-09565-2

López-Silva, P., & Madrid, R. (2024). ¿Es la noción de 'neuroderecho' un concepto jurídico indeterminado?. *Lus et Praxis, 30*(3). https://doi.org/10.4067/S0718-00122024000300004

López-Silva, P., Valera, L. (2022). Neurotechnologies and the human image: open questions on neuroprotection. In P. López-Silva, L. Valera, (Eds.), *Protecting the mind. Ethics of science and technology assessment* (Vol. 49). Springer. https://doi.org/10.1007/978-3-030-94032-4_15

López-Silva, P., & Valera, L. (2022). Protecting the mind: Challenges in law, neuroprotection, and neurorights. Amsterdam: Springer.

López-Silva, P., Wajnerman-Paz, A., & Molnar-Gabor, F. (2024). Neurotechnologies and the problem about how to protect mental privacy. *Neuroethics, 17*, 31. https://doi.org/10.1007/s12152-024-09565-2

Muñoz, J. M., & Marinaro, J. Á. (2023). Neurorights as reconceptualized human rights. *Front. Polit. Sci., 5*, 1322922. https://doi.org/10.3389/fpos.2023.1322922

Ngai, J. (2024). BRAIN @ 10: A decade of innovation. *Neuron, 112*(18), 3003–3006. https://doi.org/10.1016/j.neuron.2024.09.007

Possa, A., & Schertel Mendes, L. (2025). Neurorights as a new manifestation of the protection of the dignitiy of the human person in brazil. *Revista Brasileira De Direitos Fundamentais & Justiça, 22*(51). https://doi.org/10.30899/dfj.v22i51.1506

Presidential Commission for the Study of Bioethical Issues. (2014). Gray matters. In *Integrative approaches for neuroscience, ethics and society* (Vol. 1). Bioethics Commission.

Yuste, R., Goering, S., Arcas, B., et al. (2017). Four ethical priorities for neurotechnologies and AI. *Nature, 551*, 159–163. https://doi.org/10.1038/551159a

Yuste, R. (2019). *Everyone has the right to neuroprotection.* Columbia Neuroright Initiative. https://nri.ntc.columbia.edu/news/rafael-yuste-and-brain-hacking-everyone-has-right-neuroprotection-originally-spanish

Zuñiga-Fajuri, A., Villavicencio Miranda, L., Zaror Miralles, D., & Salas Venegas, R. (2021). Chapter Seven - Neurorights in Chile: Between neuroscience and legal science. In M. Hevia (Ed.), *Developments in neuroethics and bioethics* (Vol. 4, pp. 165–179). Regulating Neuroscience: Transnational Legal Challenges. https://doi.org/10.1016/bs.dnb.2021.06.001

Conceptual Lessons and Applications

Free Will Skepticism, Neurorights, and Coercive Moral Neuroenhancements: An Approach from Neuroabolitionism

Diego Borbón

Abstract This chapter explores the intersections between free will skepticism, the emerging concept of neurorights, and the ethical implications of coercive nonconsensual moral neuroenhancements. Free will skepticism in general, and hard incompatibilism in particular, questions whether individuals have alternative possibilities and ultimate control—in the degrees required for free will and moral responsibility—in a deterministic or indeterministic universe. Then, upon the denial of free will, some might suggest that implementing coercive neurointerventions to prevent and correct deviant antisocial behavior is necessary or desirable. This discussion is crucial to neurorights, which aim to protect cognitive liberty and consent. However, this chapter argues that neurorights, as currently conceived, are inadequate and could be less protective and more regressive in terms of human rights. Therefore, a skeptical or a hard incompatibilist approach to free will should reconsider the debate on moral enhancements and neurorights, advocating for more humane perspectives aligned with penal neuroabolitionism, which advocates for the gradual abolition of criminal law and retributive punishment, while respecting human dignity, freedom of thought, bodily integrity, autonomy, and consent. This chapter rethinks future public criminal policies and radically challenges the way in which we currently fail to do justice.

Keywords Neurorights · Hard incompatibilism · Cognitive liberty · Penal abolitionism · Quarantine model · Moral enhancement · Neurocorrections · Neurointerventions

---

D. Borbón (✉)
Center for Studies on Genetics and Law, Research Group on Biological Sciences and Law, Universidad Externado de Colombia, Bogotá, Colombia
e-mail: diego.borbon1@uexternado.edu.co

Institute of Biomedical Ethics and History of Medicine (IBME), University of Zurich, Zurich, Switzerland

P. López-Silva (ed.), *Contextualizing Neuroprotection: Latin American Perspectives on the Impact of Neurotechnological Development in Life and Society*, Ethics of Science and Technology Assessment 54, https://doi.org/10.1007/978-3-031-96055-0_2

1 Introduction

The intersections between free will skepticism, neurorights, and moral neuroenhancements introduce critical questions for both philosophical and legal domains. The rapid advancements in neurotechnologies, particularly those enabling potential "moral" neuroenhancements, pose new ethical and human rights challenges. These issues are most pertinent when viewed through the lens of free will skepticism—particularly hard incompatibilism—which questions whether humans have the kind of free will required for moral responsibility. Hard incompatibilism argues that neither determinism nor indeterminism allows for the kind of ultimate control and alternative possibilities necessary for genuine moral responsibility, leaving significant implications for how society addresses deviant behavior, particularly through neurotechnological interventions.

Free will skepticism challenges the foundational assumptions of moral and legal responsibility, raising concerns about the legitimacy of punishment, especially in criminal law. If individuals lack the kind of ultimate control or alternative possibilities necessary for moral responsibility, as hard incompatibilists argue, then punishing individuals based on their actions becomes questionable. This skeptic perspective has led some to propose alternative frameworks, such as the public health-quarantine model of Caruso & Pereboom (2022), or a more radical approach towards the complete abolition of criminal law, known as penal neuroabolitionism (Borbón, 2022, 2024). Instead of punishing individuals, a society rooted in hard incompatibilism might focus on social justice with preventive and restorative approaches, aiming to reduce harm without recourse to a criminal justice system.

Simultaneously, the future rise of neurotechnologies hypothetically capable of enhancing or altering moral capacities, often referred to as "moral" neuroenhancements, presents a new domain for ethical concern. These technologies could, in theory, be used to modify individuals' behavior to prevent and "fix" antisocial or deviant conduct. However, this raises significant concerns about autonomy, consent, and the potential for coercive use by states or institutions. Coercive and nonconsensual moral neuroenhancements could violate what scholars like Sententia (2006) and Bublitz (2013) define as "cognitive liberty"—a right to mental self-determination and freedom of thought that implies also a right to refuse neurotechnological interventions.

Neurorights, as a novel concept developed to safeguard individuals against unethical uses of neurotechnologies, aim to protect rights such as cognitive liberty, mental integrity, and privacy. However, the current relative framing of these rights remains controversial and may not be sufficient to protect against coercive neurointerventions (Díaz Soto & Borbón, 2022). While these new neuro-rights seem like promising steps forward, its implementation remains fraught with legal uncertainties, particularly when balanced against state interests such as public safety or crime prevention, as current proposed neurorights are usually understood as being of a relative nature.

In light of these challenges, this chapter argues that free will skepticism, particularly hard incompatibilism, provides a critical perspective for reconsidering the relationship between neurotechnologies and human rights. By acknowledging that individuals may lack moral responsibility, we can better address the ethical issues surrounding coercive moral neuroenhancements. The chapter will also propose that *neuroabolitionism* offers a more humane and ethically consistent framework for addressing deviant behavior, one that respects human dignity and autonomy in ways that coercive neurointerventions cannot. Instead of using neuroscience to perpetuate structural injustices with the use of coercive interventions on the name of public safety; neuroscience can be a useful tool to rethink justice based on reflections on the problem of free will, the harms of incarceration, and better public health and social justice approaches.

This chapter will proceed by first outlining the key tenets of free will skepticism and hard incompatibilism, followed by an examination of the ethical implications of moral neuroenhancements. It will then critically assess the concept of neurorights, highlighting their limitations in protecting freedom of thought, bodily integrity, consent, and human dignity. Finally, it will propose neuroabolitionism as a preferable alternative, advocating for a non-reductionist proposal focused on addressing current structural injustices, with social policies and mental health approaches, where neurotechnologies, instead of being used against people´s consent, are limited to clinical consensual interventions in exceptional cases, by prioritizing social solutions first.

2 Free Will Skepticism and Hard Incompatibilism

The debate on free will and determinism is an enduring issue in both philosophy and science. At its core, determinism implies that every event, including human actions, is the inevitable result of preceding causes governed by natural laws of physics. This view is often associated with classical mechanics, where the universe operates like a machine, unfolding according to predictable laws. In the past, Democritus, Spinoza, or Pierre-Simon Laplace, have supported some sort of this causal view. Laplace famously argued that if a sufficiently intelligent entity could know the state of every particle and force in the universe at a given time, it could know the entire past and future (Laplace, 1951).

The deterministic perspective on human action, particularly in the context of moral responsibility, posits that if our actions are governed by causal-deterministic laws, then it becomes questionable whether individuals can be morally responsible for their actions. Pereboom (2001), a key figure in free will skepticism, advocates for *hard incompatibilism*, which suggests that neither determinism nor indeterminism provides room for the type of free will required for moral responsibility. According to Pereboom, even if determinism is false and some indeterministic processes influence our decisions, such randomness does not grant us the control required for moral responsibility.

This philosophical stance aligns with neuroscientific findings that emphasize the brain's role in decision-making, which precedes conscious awareness. In experiments by Libet et al. (1983), it was shown that brain activity related to making decisions occurs before individuals report being aware of their decision. This temporal gap raises doubts about the extent to which conscious intentions guide actions, further complicating traditional notions of free will. These Libet-type experiments have received extensive criticisms. Trevena and Miller (2010), for their part, argued that Libet's findings failed to demonstrate that voluntary movement was initiated unconsciously. Guggisberg et al., (2013), Glannon (2011), and Mele (2014) also argue against the methodology and conclusions of these experiments.

In that sense, it seems more plausible to extend the arguments against free will with a different approach to neuroscience. The brain and the nervous system are composed of billions of neurons, glial cells, cerebrospinal fluid, blood, among others. Like any other part of the human body, our brains fundamentally operate by physical, electrical, and chemical signals that are way beyond human awareness and causal-agent control (in a *causa-sui* sense). In other words: brains and bodies operate by automatic processes in interaction with the environment, and we cannot cause our own cause because we cannot break the causal chains of the laws of physics and the remote past. Then, a materialistic approach to consciousness may account for the intuitive and folk psychological belief in "free will" (and consciousness too) as a phenomenon of deterministic mechanisms of the body. Free will, then, is a utilitarian illusion and socially situated belief that, in reality, cannot metaphysically be justified.

In line with these perspectives, deterministic models such as those in classical mechanics suggest a fixed universe where everything is predetermined, leaving no room for moral responsibility. As Newtonian physics dictates, the second law of motion implies that objects move based on forces acting upon them (Feldman, 2002), leading us to argue that human actions (but also the social and cultural environment) are determined as human bodies (and the external reality itself) are fundamentally bind to material laws of physics. Furthermore, Harris (2012) succinctly captures this perspective, stating that "determinism, in any relevant sense for human behavior, is true" (8). He emphasizes that the neurological events underlying our decisions are dictated by prior causes beyond our control, and even that the *self* is an illusion. What we call "the self" would then be merely a constantly changing stream of thoughts and sensations generated by the material brain, and no stable or separate subject remains apart from it.

In my view, the set of physical laws that govern the material universe, in turn, rejects the possibility that a person can be *causa sui*, that is, its own free cause. Consequently, for human beings to escape the truths of causal determinism, we would need to have the ability to break the chain of preceding causes and be capable of causing our own cause. However, for such a thing to be possible, the intervention of some immaterial substance not subject to the deterministic laws of physics would be necessary, akin to the mystical idea of the soul, a thesis we reserve for theology, but not science. If there is no *ghost behind the machine*, no little *homunculus* inside our brain that is the *true self*, or if our *soul* is not located in the pineal gland, as Descartes claimed, then, of course, we must reject the possibility that human beings possess a mystical power that allows us to cause our own cause.

However, the conversation becomes more complex when quantum mechanics is introduced, as we then face the challenges of indeterminism at the subatomic level. Quantum effects, such as the superposition, the uncertainty principle, the wave-particle duality, and the collapse of the wave function, challenge strict deterministic models. While indeterminism may introduce an element of unpredictability, it does not grant individuals the control needed for free will and moral responsibility, as random events impede any form of moral imputability: we cannot be responsible for billions of quantum indeterministic operations at the subatomic level. If our actions are merely product of random quantum chance, we then lack ultimate control over decisions needed for moral responsibility.

Also, scholars have challenged the traditional indeterministic interpretations of quantum mechanics. For example, according to Hossenfelder and Palmer, *superdeterminism* refers to a class of theories that resolve the measurement problem in quantum mechanics by rejecting the usual assumption of "Statistical Independence", which states that the choice of measurement settings is independent of the hidden variables describing a quantum system (Hossenfelder & Palmer, 2020). By allowing the measurement settings to be correlated—via an ordinary local process—with the hidden variables of the particles themselves, superdeterministic models can reproduce the observed quantum predictions without resorting to nonlocality or genuine randomness. In this view, what might appear to be a "conspiracy" between detector settings and particle properties is, instead, a physically consistent violation of Statistical Independence. Importantly, *superdeterminism* is not a mere interpretation of quantum mechanics but rather a proposal for a more fundamental, deterministic theory from which standard quantum mechanics would emerge.

In simpler words, *superdeterminism* is the idea that how we choose our measurements in a quantum experiment is not truly independent from the properties of the particles, so everything (including our choices) is linked in one grand, deterministic system. One can also interpret that "quantum chance" arises more from our limited knowledge—an epistemological gap—than from any inherent indeterminism in the atomic or subatomic particles themselves, meaning that a particle is not truly in all places at once, but appears so only until we measure and reveal its actual deterministic state. Chance in quantum mechanics might be an epistemological problem (a gap of knowledge), not a true ontological property (a state of the particle itself).

In any case, the implications of free will skepticism are profound for both legal and moral responsibility, as they challenge the foundational assumptions of retributive justice. Retributivism, a dominant theory in many criminal justice systems, is grounded in the idea that individuals deserve to be punished because they freely choose to commit wrongful acts. However, free will skepticism, particularly as advocated by Derk Pereboom and Gregg Caruso, undermines this notion by arguing that individuals do not have the free will over their actions that would make them deserve blame or punishment.

Pereboom (2001) proposes the concept of "hard incompatibilism," which suggests that neither determinism nor indeterminism allows for the type of free will necessary for moral responsibility. If determinism is true, our actions are the inevitable outcome of prior causes; if indeterminism is true, our actions are influenced by random factors, which still do not grant us control over them. This leaves little room for the desert-based notion of responsibility that retributivism relies upon. Instead, Caruso & Pereboom (2022) advocate for a quarantine model, which parallels how society treats individuals who are a threat to public health. Like carriers of a contagious disease, according to Caruso & Pereboom (2022), offenders should be prevented from causing harm—not because they deserve punishment, but because society needs protection. The quarantine model calls for preventive quarantine incapacitation without invoking moral blame, emphasizing rehabilitation and the well-being of the offender rather than punishment for retribution (Caruso & Pereboom, 2022).

Gregg Caruso builds on this model with his public health-quarantine approach, which integrates public health ethics into criminal justice. Caruso (2021) argues that society should focus on addressing the determinants of criminal behavior to prevent crime rather than punishing offenders after the fact. His model promotes preventive measures over punitive responses. By viewing criminal behavior through the lens of public health, this approach seeks to shift the focus from retribution to prevention and rehabilitation, aligning closely with the perspective of free will skepticism.

Both Pereboom and Caruso agree that the traditional concept of moral responsibility, grounded in the idea of basic moral desert—the notion that people deserve to be praised or blamed simply for their actions—is flawed. They argue that by rejecting retributivism and embracing a model focused on incapacitation and prevention, society can adopt a more humane approach to criminal justice. This shift has significant implications for how we think about legal responsibility. If individuals are not truly free agents, then the justification for retributive punishment falls apart, and alternative models like quarantine, which prioritize rehabilitation and societal protection, become more appealing.

Hard incompatibilism, as defended by Derk Pereboom and Gregg Caruso, fundamentally challenges our understanding of moral responsibility. If we accept the premise that individuals lack free will, either because actions are fully determined by prior causes or because indeterministic events cannot grant moral responsibility, then the notion of holding individuals accountable for their actions in the traditional sense becomes nonsensical. This might also have implications for "neuroenhancements", particularly when these technologies are framed as tools for correcting deviant behavior.

From a standpoint that denies free will, if individuals are not morally responsible for their actions, whether due to deterministic or indeterministic processes, then it might become ethically questionable to subject them to coercive moral neuroenhancements designed to "fix" their deviant determined behavior. This might be argued if the very foundation of retributive justice collapses under free will skepticism, suggesting that, instead of punishment, then interventions and corrections might be reconsidered in favor of non-retributive, preventative, or rehabilitative models.

If we are not the ultimate authors of our actions, the concept of moral enhancement takes on a different ethical dimension. Rather than focusing on incarceration for antisocial behavior, the ethical framework might instead emphasize biomedical interventions. These types of interventions, particularly those that might be coercive and without informed consent, would violate an individual's autonomy and cognitive liberty—principles that are increasingly being recognized in neuroethical discussions.

This leads to an essential transition point: while free will skepticism provides significant ethical reasons to question the justifiability of retributive punishment, there might also be arguments that some forms of neurointerventions could be seen as necessary, even without consent, to prevent ultimate harm and antisocial behavior. If individuals are not fully in control of their actions, might it be justifiable to use neuroenhancements to prevent harm or promote social well-being? This would shift the conversation towards utilitarian justifications of behavior modification, which may conflict with the individual's right to cognitive liberty, freedom of thought, dignity, informed consent, bodily integrity, and autonomy.

Thus, free will skepticism poses critical questions for the ethical justification of neuroenhancements, particularly when considering the balance between social safety and individual rights. This analysis leads us into the next sections, which examines the ethical concerns surrounding neuroenhancements and whether an appropriate framework can be developed to protect individuals' rights while addressing societal concerns.

3 Moral Neuroenhancements and Its Implications

According to Persson and Savulescu (2012) moral bioenhancement involves biomedical interventions to promote a more altruistic and justice-oriented behavioral profile, including the reduction of immmoral emotions or dispositions. In that sense, these "enhancements" can be understood as interventions or corrections that might involve biological interventions designed to improve moral decision-making processes by targeting specific biological mechanisms. These enhancements can theoretically promote prosocial behaviors such as empathy, fairness, or altruism, while reducing antisocial tendencies like aggression, impulsivity, or selfishness. In recent years, scholars have debated the ethical implications of using such technologies, especially in contexts of criminal or harmful behavior. It might be argued that moral enhancements offer a alternative that might potentially replace incarceration and even criminal sanctions with interventions aimed at directly altering the neurobiological underpinnings of deviant behavior.

An initial paper by Persson and Savulescu (2008) argues that the rapid advancement of technology and cognitive power increases the potential for catastrophic harm to access and misuse destructive capabilities such as biological, nuclear, or other mass destruction weapons. While cognitive enhancement could accelerate scientific progress and bring societal benefits, Persson and Savulescu (2008) argue that it also heightens risks if not accompanied by significant "moral enhancement". In that sense, they argue that if safe moral enhancements are ever developed, not necessarily by neurotechnological means, these could be obligatory and compulsory (Persson & Savulescu, 2008).

In their book "Unfit for the Future: The Need for Moral Enhancement" Persson and Savulescu (2012) argue that rapid scientific and technological advancements have transformed human living conditions, creating global and long-term impacts, including environmental degradation, climate change, and the risk of mass destruction through nuclear and biological weapons. In that sense, Persson and Savulescu (2012) suggest that liberal democracies face significant challenges necessitating the moral enhancement of citizens to expand their concern to future generations and global issues. They maintain that traditional moral education has proven insufficient, suggesting the potential need for biomedical moral enhancement, which could influence central dispositions like altruism and justice.

A year later, in a very brief response to some critiques, Persson and Savulescu (2013) argue that while many societal restrictions on freedom are involuntary, they can be justified if they prevent ultimate harm. They suggest that not imposing compulsory moral bioenhancement could lead to catastrophic consequences, making life on Earth impossible if harmful acts are not prevented. Persson and Savulescu emphasize the necessity of preventing harm before it occurs. They acknowledge that while incentives for voluntary enhancement could be effective, there may be scenarios where compulsory enhancement is morally justified to ensure safety. At the end, they conclude that the current research in moral bioenhancement is too nascent for them to decide whether it should be obligatory or voluntary (Persson & Savulescu, 2013).

De Grazia (2014) also explored the concept of moral bioenhancement as a means to improve human moral behavior through biomedical interventions, arguing that it merits serious consideration due to the possible inadequacies of traditional moral improvement methods in addressing pressing global moral challenges. He acknowledges the ethical complexities, especially within the context of moral pluralism, and suggests that any state-supported enhancement initiatives should align with widely agreed-upon moral standards. De Grazia (2014) maintains that, if safe and effective, it could be morally desirable without inherently compromising human freedom, which, while important, is not the only value to prioritize in moral behavior.

De Grazia (2014) then challenges concerns raised by critics like John Harris, who argue that bioenhancement would undermine freedom and thus the intrinsic value of moral behavior. By distinguishing between different types of freedom and showing that compatibilist views of freedom align with enhancements, DeGrazia argues that those interventions do not necessarily pose an inherent threat to human agency. He suggests that while freedom is valuable, moral behavior itself and its beneficial outcomes are also crucial. The paper concludes that moral bioenhancements could contribute to significant moral progress without eliminating human freedom and should be considered as a potential tool.

It is interesting to see that in 2008, Thomas Douglas, in his paper "Moral enhancement" argued against the Bioconservative Thesis, which posits that biomedical enhancement, even if feasible and legally permissible, would not be morally permissible. Douglas (2008) presents moral enhancement as a counterexample, suggesting that biomedical interventions aimed at improving an individual's motives—such as attenuating emotions like racial aversion or violent aggression—could be morally permissible and beneficial. He explains that these enhancements would not only likely lead to better overall motives without disadvantaging others but would also align with moral improvement across plausible ethical theories.

Douglas (2008) anticipates several objections, including concerns about identity change, freedom restriction, and unnatural means. He addresses these by arguing that moral enhancement would likely alter an individual's "brute self" to free the "true self," thus increasing rather than restricting autonomy. He also challenges the notion that moral enhancement by biomedical means is inherently unnatural or objectionable, highlighting that moral reasons and consequences justify such self-improvement. In cases like Smith's hypothetical scenario, where a person (Smith) chooses moral enhancement to mitigate counter-moral emotions, Douglas concludes that it would be morally permissible and could counter bioconservative arguments, though he acknowledges the need to assess broader societal implications and potential misuse of such technologies.

In sum, it is clear that the application of moral neuroenhancements raises profound ethical questions. Most scholars in favor of these types of interventions are even skeptic on the nonconsensual application. In that sense, one critical concern is the issue of informed consent, particularly when such interventions are proposed as alternatives to punitive measures like imprisonment and if they are applied in a coercive matter. This introduces a dilemma: while the goal of reducing harm is laudable, overriding individual autonomy to achieve this goal may infringe on fundamental rights. Moreover, the line between consensual enhancements and coercion becomes blurred when the state or other authorities mandate neurointerventions appealing to "rational choice" as an alternative to induce consent: either you consent to the intervention or you might be subject to harsh alternatives, such as long incarceration periods or even coercive intervention in absence of voluntary commitment.

From a philosophical standpoint, free will skepticism, as discussed earlier in this chapter, further complicates the justification for moral neuroenhancements. If individuals are not morally responsible for their actions due to deterministic processes beyond their control, the ethics of altering their moral capacities become ambiguous. On one hand, the deterministic view might suggest that individuals cannot be blamed for their actions, but on the other hand, this may also lead to justifying interventions aimed at preventing future social harm. If a person lacks free will, does that mean that consent is no longer important? Some could argue that in extreme cases moral enhancement could even be compulsory and this raises critical concerns about the scope and limits of such interventions, as well as the broader implications for human rights.

In the paper *Criminal Rehabilitation Through Medical Intervention*, Douglas (2014) also explores the ethical challenges of requiring criminal offenders to undergo medical interventions aimed at rehabilitation. Douglas questions the common assumption that such interventions should only be administered with the offender's valid consent. He argues that, in certain cases, committing a crime could render an individual morally liable to nonconsensual medical interventions. Douglas highlights the complexity of balancing rehabilitation with respect for bodily integrity and autonomy, especially when the interventions are imposed as part of criminal justice measures, such as conditions of parole or sentencing. He suggests that the right to bodily integrity, while significant, may not always preclude nonconsensual medical correctives, particularly when compared to other accepted nonconsensual punishments like incarceration, which restricts freedom of movement without consent (Douglas, 2014).

Upon these discussions, Bublitz (2016) identifies three senses of freedom that could be compromised by enhancements: freedom of action, freedom of will, and what he calls freedom of mind. He argues that advocates of those interventions often neglect the broader concept of freedom of mind, which encompasses an individual's capacity to think and make decisions free from external and internal interference. While some proponents assert that it does not threaten freedom since it promotes pro-social behavior, Bublitz contends that these interventions can subtly undermine autonomy by altering emotional and cognitive processes in ways that bypass rational deliberation. This raises critical concerns about the ethical boundaries of using biomedical means to enforce moral behavior.

Bublitz (2016) critiques the argument that bioenhancements particularly those targeting emotions to encourage empathy or reduce aggression, should be considered benign because they yield morally desirable outcomes. He points out that altering emotional states through biochemical means may not enhance true moral agency but rather shift decision-making toward instinctual responses that compromise conscious control and rational thought. This form of intervention, Bublitz (2016) asserts, can lead to a diminished capacity for mental freedom, where individuals become subjects of external manipulation rather than autonomous moral agents.

Lastly, Bublitz (2016) highlights the political and legal ramifications of neglecting freedom of mind. He warns that the failure to recognize this type of freedom could pave the way for state overreach, where governments justify interventions that shape citizens' thoughts and beliefs under the guise of promoting social order. Drawing parallels with dystopian scenarios, Bublitz (2016) emphasizes that preserving freedom of mind is crucial to maintaining a boundary between liberal democratic principles and authoritarian control. In that sense, Bublitz (2016) concluded that "[w]ithout engaging with freedom of mind and the limits of state power, their proposals are premature and dangerously off-point" (103).

To sum up, moral bioenhancement entails interventions on the biological functions, including the nervous system or related genetic mechanisms, to hypothetically promote more altruistic and justice-oriented behaviors. These interventions aim to modify bodily processes to foster prosocial attributes such as empathy and fairness while mitigating antisocial impulses like aggression. The hypothetical potential benefits of such "enhancements", particularly in the context of criminal behavior, might include reducing recidivism and fostering rehabilitation. However, the possibility that these interventions could be coercive and nonconsensual introduces significant ethical and human rights concerns. Such coercive interventions risk undermining individual autonomy, raising the stakes for a debate on the boundaries between public safety and individual fundamental rights. The ethical dilemma is compounded when considering the extent to which informed consent and cognitive liberty are safeguarded within coercive or state-mandated interventions.

Then, the implications of coercive neuroenhancements are profound, especially when viewed through the lens of human dignity, freedom of thought, autonomy, and bodily integrity. Such interventions challenge the core ethical principles that uphold those human rights, as imposing changes on an individual's neural processes without their consent compromises their fundamental rights. The right to freedom of thought—a cornerstone of democratic societies—risks being diminished when external forces can alter an individual's cognitive processes to conform to stereotypical-socially approved moral norms. This breach of mental sovereignty not only impacts personal autonomy but also the broader social fabric that relies on the protection of these rights to foster diversity of thought and moral pluralism.

Freedom of thought, as it pertains to law, human rights, and neuroethics, has been increasingly recognized as a critical aspect of protecting mental autonomy and personal identity. In academic discussions, freedom of thought is often linked to the right to maintain one's mental processes, beliefs, and cognitive patterns without unwarranted interference. This principle is foundational to preserving human dignity and individuality in the face of advancing neurotechnological capabilities. Neuroenhancements, particularly those applied coercively, threaten to undermine this freedom by altering the neural substrates that underpin conscious thought and decision-making. Such interventions challenge the notion of an individual's mental domain being inviolable, raising questions about the extent to which the state or institutions can ethically justify tampering with cognitive processes for perceived social benefits. I will come back to these discussions in a posterior section, also by establishing some conceptual differences between free will and freedom rights, natural freedom or basic notions of positive freedom. I will suggest that denying free will is perfectly compatible with advocating in favor of freedom rights, consent, and human dignity.

Moreover, the violation of bodily integrity becomes a pressing issue when considering the physical nature of neuroenhancements. Unlike conventional punitive measures, which primarily restrict freedom of movement or impose social consequences, coercive neuroenhancements intrude directly into the body's neurobiological systems and the mind, which is the most private and intimate aspect of a human being. This level of intervention amplifies concerns about informed consent, as it effectively transforms the body into an object of external manipulation. The ethical tension between the potential societal benefits of hypothetical reduced criminal behavior and the preservation of individual rights highlights the need for a robust framework that prevents such enhancements from undermining personal dignity.

Bublitz (2016) underscores that while proponents argue for the beneficial outcomes of moral neuroenhancements, they often overlook the broader implications on the freedom of the mind. Unlike freedom of action or will, freedom of mind is more subtle but integral, protecting against both overt and covert manipulation. This framework sets the stage for understanding how current conceptions of neurorights may fall short. These rights, despite their forward-looking aims, remain overly relative and potentially inadequate in preventing state overreach or the erosion of cognitive autonomy. Moving forward, the following section will examine these limitations in-depth, showing how neurorights, as currently framed, may not offer robust protection and instead could become tools that enable rather than prevent coercive practices. In that sense, neurorights are not a progressive proposal but rather a regressive one as they are less protective than current human rights standards of absolute protections on human dignity and freedom of thought. We will advocate for an interpretation of these rights as absolute, regarding external nonconsensual state interference in criminal law.

4 Neurorights as Problematic Relative Protections

In light of developments in human neuroscience and neurotechnology, the formal proposal to establish a new category of fundamental rights known as 'neurorights' emerged in 2017. The seminal work by Ienca and Andorno (2017) marks a pivotal starting point in this field. According to Ienca (2021), the conceptual foundation for neurorights began in the early 2000s with the work of neuroethicists like Boire and Sententia on cognitive liberty, which they argued should be regarded as a fundamental right, integral to broader freedoms.

Ienca and Andorno (2017) posited that the growing sensitivity and accessibility of neurotechnological devices would necessitate the creation of four new rights. They suggested that human rights have historically emerged as responses to threats against fundamental human interests, and the same process might unfold with neurotechnology. Accordingly, they proposed four neurorights: the right to cognitive liberty, the right to mental privacy, the right to mental integrity, and the right to psychological continuity.

For Ienca and Andorno (2017), the right to cognitive liberty protects individuals as a negative right, which means the possibility to refuse interventions. The right to mental privacy safeguards private or sensitive mental information from unauthorized collection, storage, usage, or deletion. The right to mental integrity focuses on protecting individuals from unauthorized interventions in their mental well-being through neurotechnological means, especially if these interventions result in physical or psychological harm. Lastly, the right to psychological continuity ensures the protection of personal identity and behavioral coherence against modifications by third parties.

In the same year, a team led by Spanish neurobiologist Rafael Yuste published a brief commentary in *Nature*, highlighting the ethical challenges posed by neurotechnology and artificial intelligence (AI), emphasizing the need to respect and preserve privacy, identity, agency, and equality. Yuste et al., (2017) noted that while current brain-computer interface (BCI) technology primarily targets therapeutic outcomes, such as aiding individuals with spinal cord injuries, advancements were moving towards deeper interpretation of neural activity and potential direct brain manipulation.

Yuste et al., (2017) argued that although these technologies could revolutionize treatment for various conditions and enhance human experience, "the technology could also exacerbate social inequalities and offer corporations, hackers, governments, or other entities new means to exploit and manipulate people" (160). They further warned that these advancements could profoundly alter essential human characteristics, such as "private mental life, individual agency, and the understanding of individuals as beings unified by their bodies" (160).

To address these concerns, Yuste et al., (2017) proposed that international and national guidelines should be established to set boundaries for enhancement technologies and define their acceptable contexts of use. To mitigate this, they recommended instituting countermeasures against biases as a standard practice, ensuring marginalized groups are involved in the design of algorithms and devices. In support of these measures, Yuste et al., (2017) concluded that "clauses protecting such rights, or 'neurorights,' should be incorporated into international treaties, such as the Universal Declaration of Human Rights" (162).

Two years after, the NeuroRights Initiative was founded at Columbia University to act as a human rights organization dedicated to creating guidelines for neurotechnological innovation. This evolved into the NeuroRights Foundation. The initiative proposed five key neurorights: Mental Privacy, ensuring that any neurodata collected must remain private and subject to deletion upon request; Personal Identity, protecting individuals from technology disrupting their sense of self; Free Will, ensuring individuals retain control over their decisions without hidden external manipulation; Fair Access to Mental Augmentation, advocating for guidelines to regulate mental enhancement technologies and uphold equitable access; and Protection Against Bias, mandating measures to counter algorithmic biases and ensure marginalized groups contribute to the design of neurotechnological tools.

Although important, neurorights have received extensive academic criticism. Christoph Bublitz (2022) stands out for his firm opposition, arguing that the creation of new neurorights risks inflating human rights to the point of diminishing the value of existing ones. He maintains that instead of generating novel rights, current human rights frameworks should be adapted to meet the challenges posed by neurotechnology. This approach, he suggests, ensures that human rights remain meaningful and do not devolve into a proliferation of categories that complicate their implementation and undermine their universality.

Bublitz (2022) also highlights the blurred lines between scientific advocacy and legislative precision in the neurorights movement. Bublitz asserts that the scientific expertise of the initiative´s leaders in neuroscience does not translate into the legislative and legal expertise required to create robust human rights laws. This disconnect, he argues, results in rights that are poorly defined, conceptually flawed, and detached from the nuanced realities of legislative practices. The rights proposed often reflect a neuro-centric bias that may not align with the broader, well-established principles of human rights law.

Adding to this critical discourse, Fins (2022) brings attention to the practical implications of enshrining neurorights in law, particularly through the lens of Chile's constitutional amendment. Fins warns that the premature adoption of such rights could hinder research and clinical practices, especially in treating patients with disorders of consciousness. The potential conflict between stringent protections and beneficial medical advancements raises concerns that overly cautious legal measures may inadvertently impede progress. Fins concludes that neurorights initiatives are premature and should not be adopted in any jurisdiction.

Other scholars echo these apprehensions. Ruiz et al., (2021) question whether the proposed rights overlap with existing legal protections, emphasizing the ambiguity and potential redundancy in their definitions. Moreu Carbonell (2021) argues that the creation of specific neurorights is not justified, advocating instead for the adaptation of current rights to new technological realities. Meanwhile, Herrera-Ferrá et al., (2022) stress the need for culturally and contextually informed frameworks, highlighting that the global discourse on neurorights may overlook diverse perspectives. These critiques collectively call for more rigorous debate and multidisciplinary dialogue before adopting any new legal categories in this evolving field.

In this sense, the principle of progressivity in human rights mandates the irreversible expansion and non-reduction of protections, as acknowledged in the preamble of the Universal Declaration of Human Rights and upheld by the Inter-American Human Rights System (IACHR, 1993). This principle ensures that advancements in human rights cannot be rolled back, emphasizing their expansion to offer greater protection (Cunego, 2016). Applying this principle to the creation of neurorights raises critical concerns. While the foundational intent behind neurorights should be to enhance the protection of individuals in the face of advancing neurotechnologies, the proposals, as currently framed, risk undermining this progressivity.

Ienca and Andorno's (2017) seminal work suggests the establishment of rights such as mental privacy, mental integrity, or cognitive liberty. Yet, they acknowledge that neurorights, akin to existing fundamental rights, possess a relative nature and can be limited under certain conditions. Specifically, they argue that the right to mental privacy could be overridden for public interest purposes, allowing the collection and disclosure of private neural data (Ienca & Andorno, 2017). Moreover, the authors imply that nonconsensual brain scans might be justifiable in particular scenarios, such as with a judicial warrant (Ienca & Andorno, 2017). This raises the contentious issue of applying neurotechnology in criminal contexts such as coercive neuroenhancements, which could profoundly impact individual autonomy and rights.

Ienca and Andorno (2017) even argue, regarding the right to psychological continuity, "that some neurotechnologically-induced personality changes could be tolerated with regard to persistent violent offenders (for instance, serial rapists, killers and pedophiles)" (23). This is said in the name of public safety and protection to society if said "dangerous individuals" are likely to reoffend, so intervening would be justified. It is very interesting so see that Ienca and Andorno even frame nonconsensual neurointerventions as a positive alternative from incarceration: "This would even be a good alternative for those individuals themselves, who could avoid in this way spending their whole lives in prison" (23). However, Ienca and Andorno (2017) conclude that before allowing these types of interventions, there is a need for caution and broader public discussions.

In my opinion, while it is true that most human rights are inherently relative and often balanced against other rights and principles, it is crucial to reinforce the non-negotiable principle of human dignity and freedom of thought as bases for absolute ban against nonconsensual state interventions. I advocate for absolute prohibitions when neurotechnology affects the essential nucleus of rights such as human dignity, freedom of thought, and bodily integrity. The use of neurotechnological tools to modify the neural correlates of personal thoughts and emotions without consent constitutes an intolerable intrusion in human rights, and its application for sentencing and coercive correction purposes amplifies these violations. In such cases, individuals' risk being reduced to mere instruments within the judicial process, raising the ethical question of whether potential offenders should be compelled to undergo a forced medical intervention in the name of public safety.

In that sense, neurorights as relative rights, may paradoxically erode protections rather than strengthen them, violating the principles of progressivity and non-regression. Instead of ensuring greater safeguards, these new neurorights risk becoming vague instruments that may allow the expansion of state punitive powers. Then the question is: do we really need new lists of relative, vague, and weak rights? Or rather, do we need some absolute protection on the essential nucleus of fundamental rights such as human dignity, freedom of thought, and bodily integrity? Law is the result of political decisions, so this is, naturally, a political decision on what the future of societies and justice systems should be like.

5 What Neuroscience Should Really Be for

As I see it, we have two big political decisions to make: either we use neuroscience as an instrument for punitive state powers, applying coercive neurocorrections in the name of public safety, or we rather use neuroscience findings to rethink law and justice. I think we are better off in the future with the second option, and penal abolitionism might help inform this alternative.

Sociological penal abolitionism, as advocated by Nils Christie, Thomas Mathiesen, and Louk Hulsman, begins by conceiving crime as a social phenomenon rather than a purely legal or moral one. These thinkers argue that criminal law primarily inflicts "sterile pain," labeling individuals as offenders or criminals while neglecting deeper social roots of conflict (Christie, 1981). In this view, prisons are ineffective, inhumane institutions that profit from pain and primarily target the most vulnerable members of society, creating a "prison industry" rather than a genuine mechanism of reintegration. The problem, abolitionists claim, is that while criminal law formally aims to reduce harm, it instead imposes institutionalized suffering—what Christie (1981, 20) terms the "conscious inflicting of pain".

Despite attempts to justify punishment as a means to limit private vengeance (Ferrajoli, 1995), many authors find that imprisonment fails in both prevention and retribution (Cid, 2009; Cullen et al., 2011; Davis, 2003; Mathiesen, 2003, 2006, 2015). In countries such as Colombia, constitutional rulings (T-153/1998, T-388/

2013, T-762/2015, SU-122/2022) have repeatedly declared an unconstitutional state of affairs in the prison system due to overcrowding, inadequate health services, and severe rights violations, effectively meaning that constitutional guarantees do not operate behind bars. Globally, over eleven million individuals are incarcerated (Fair & Walmsley, 2021), yet evidence shows that the aims of imprisonment are seldom achieved, and conditions often amount to cruel, inhuman, or degrading treatment.

In this context, the emerging insights of neuroscience and neuropsychology provide powerful additional arguments against the overreliance on incarceration. Numerous studies link imprisonment to deteriorating mental health and adverse cognitive outcomes (Borbón, 2022; Brinkley-Rubinstein, 2013; Constantino et al., 2016; Edgemon & Clay-Warner, 2019; Haney, 2012, 2017; Huey & McNulty, 2005; Meijers et al., 2015, 2018; Nurse et al., 2003; Piper & Berle, 2019; Reiter et al., 2020; Schnittker et al., 2012). From an abolitionist perspective, applying a scientific understanding of human behavior—rather than persisting with punitive, pain-driven approaches—could foster more restorative, equitable, and effective methods of dealing with conflict.

Most public criminal policies in Latin America are shaped by irrational and reactionary penal approaches, driven more by public opinion than by genuine criminological research. These policies are further aggravated by their selective nature, as the punitive power of the state disproportionately targets vulnerable populations. In Colombia, for example, the National Penitentiary and Prison Institute INPEC (2025) reports that 96.34% of incarcerated individuals are illiterate, have only attended primary school, or completed high school, while only 297 out of 103.885 incarcerated people have received any type of postgraduate education.

As sociologist Wacquant (2009) observes, prisons function as warehouses for the poor and those deemed undesirable by society. Also, at the regional level, the Inter-American Court of Human Rights (IACHR, 2017) has condemned multiple countries, including Argentina, Guatemala, Peru, Venezuela, and Ecuador, for deplorable prison conditions, such as overcrowding, poor sanitation, inadequate food, and lack of basic necessities, alongside violations of physical and mental integrity, cruel treatment, and sexual violence. The IACHR (2017) has characterized these sanctions as forms of corporal punishment and mandated reparations, policy reforms, and investigations into those responsible.

Ultimately, prisons serve the inhuman before serving the human, causing severe neuropsychological harm to inmates while failing to achieve any purpose beyond isolating individuals from society. By incarcerating marginalized groups, the state delivers an empty symbolic message while failing to address the social structures that lead individuals to commit offenses.

Even if one disagrees with the arguments for free will skepticism, at least some of us may agree on the need to limit the punitive power of the state. Criminal law systems cause excessive harm, without transforming social structures, under the metaphysically unjustified belief in free will, and it is time to transcend this narrative.

The abolition of criminal law invites a profound reconsideration of justice, and neuroscience can play a useful role in this matter. In essence, criminal law's foundation on the questionable premise of free will, combined with the potential misapplication of neurorights, underscores the need for a shift toward policies that prioritize human dignity and comprehensive, science-based prevention measures based on social justice and public health policies. Instead of creating ambiguous new rights, the focus should be on developing existing frameworks to protect individuals against the misuse of neurotechnology and promote societal well-being. The transition from a punitive to a restorative, dignity-centered approach represents an evolution in justice, supported by neuroscientific findings that emphasize social policies, rehabilitation, and prevention over retribution.

That is why I propose a new framework towards the abolition of criminal law, based on complex thought and non-reductionist approaches to human behavior. This entails understanding scientifically both the social-environmental factors of human behavior, but also the individual and clinical elements of antisocial behavior. That is why I advocate in favor of five principles to debate on the future of justice:

First principle: Without free will, there can be no criminal law based on culpability

The principle of culpability in criminal law assumes that individuals possess free will and can therefore be held responsible for their actions. This principle hinges on the ability to act otherwise and on conscious control over decisions. However, the strong incompatibilist thesis argues that both determinism and indeterminism negate the justification on the belief of free will. If human behavior is determined by neurobiological and environmental-social-cultural factors or influenced by random events, there would be no ultimate control or alternative possibilities for action, thereby undermining the basis of culpability and, consequently, criminal law.

Second principle: The prison system generates adverse neuropsychological effects and should be abolished

Research in neuropsychology has shown that incarceration and the prison environment have detrimental effects on prisoners' mental health and brain function. Studies such as those by Meijers et al., (2018) indicate that only after three months of imprisonment, prisoners experience a decline in brain functions related to self-regulation, attention, and impulse control. These effects worsen reintegration prospects and increase the risk of recidivism. In Latin American countries, where prison conditions are substandard, penitentiary practices violate human dignity and perpetuate a cycle of suffering and social disintegration. In this debate, the public health quarantine model of Pereboom and Caruso might be justified in very exceptional cases in which social approaches, restorative justice, and mental healthcare fail. But, in my opinion, an incapacitation model should also be a last resource, as quarantines might end up replicating the same logics of prisons. I worry that the incapacitative side of the quarantine approach could end up reproducing prison-like dynamics—or, worse, echo the shortcomings and lessons learned of the United Kingdom's abandoned Dangerous and Severe Personality Disorder (DSPD) programme, which targeted individuals with high-risk personality disorders. As Tyrer et al., (2015) note, a leading criticism of the DSPD model was that it became a pretext for holding patients well past their sentence lengths, a practice critics labeled as "warehousing" (p. 102). In any case, a realistic abolitionist model must propose long-term prison replacement. In the medium term, it is essential to dignify prison environments and life in prison, as well as prioritize social policy approaches that prevent the need of incarceration itself.

Third principle: There are better alternatives to criminal law

Neuroabolitionism advocates for a gradual replacing of criminal law with justice systems that prioritize conflict resolution in a humane and dignified manner. Civil justice, restorative justice, and therapeutic jurisprudence offer models focused on resolution through reparation and well-being, involving the community and victims in the process. These models avoid the state's expropriation of conflicts and reject the brutality of retributive punishment, aiming to restore relationships and improve the psychological health of those involved. A restorative approach would also prioritize victims' rights and voices. In conflicts where social factors of marginalization and structural injustice particularly determined the offender's behavior, an environmental, social, labor, economic, restorative, and mental health response would be necessary. Similarly, in those few cases where the only explanation for the behavior is a neurobiological dysfunction, a consensual medical response would be a priority.

Fourth principle: Neuroscience, criminology, and sociology can better assess deviant behavior

Collaboration among neuroscience, criminology, and sociology can provide preventive solutions to mitigate antisocial behaviors while addressing also structural injustices. Studies have shown that factors such as extreme inequality and social exclusion leave neuropsychological impacts that may lead to maladaptive behaviors. Implementing prevention programs based on scientific evidence can reduce the need for punitive penal systems, promoting a transition to more humane and effective forms of social control. In this sense, an alternative model to criminal law should not focus on incapacitation, as does Pereboom and Caruso's public health quarantine model. Rather, a better abolitionist alternative to criminal law should focus on scientifically studying the neurobiological, environmental, social, and cultural factors that determine human behavior and the social response. Prioritizing the decriminalization of less harmful behaviors, as well as rigorous social policies to prevent harmful behaviors, focuses on prevention, not punishment nor incapacitation. Public policy priorities should focus on avoiding even the possibility of needing incapacitation.

Fifth principle: Human dignity, freedom of thought, consent, and bodily integrity must be the limit of science, technology, and law

Neuroabolitionism posits that human dignity must guide any development in justice and the use of neurotechnologies. Coercive brain interventions can only be considered "moral" if they have the informed consent of the individual, respecting the bioethical principle of autonomy. Totalitarian proposals that instrumentalize people and violate their mental privacy and integrity need to be rejected. This humanistic approach aims to protect the mind as the last space of personal freedom and opposes measures that undermine integrity and human rights.

The protection of the essential core of human rights, regarding human dignity and freedom of thought, must remain absolute when faced with coercive uses of neurotechnology for moral enhancement. Bodily integrity, informed consent and autonomy should also be protected. This unwavering protection ensures that state or institutional power cannot override individual rights in ways that could lead to significant violations. While certain irrelevant, nonbiomedical, nor invasive social interventions do not violate the core of these rights, it is critical to establish strict boundaries to prevent any form of coercion that would inflict harm on these fundamental human values, as coercive neurocorrections do.

Absolute prohibitions should be specifically directed at bioneurotechnological interventions that pose a high risk of brutality and invasiveness or even damage to the core of human rights. The application of such coercive measures without informed consent undermines the principle of human dignity, treating individuals as mere instruments of public policy rather than autonomous beings with intrinsic worth. This perspective aligns with the fundamental tenets of international human rights law, which recognizes dignity as a non-derogable right that forms the foundation for all other rights (Díaz Soto & Borbón, 2022), as freedom of thought which is generally understood as an absolute right (Bublitz, 2021).

Moreover, while state interventions may sometimes be necessary for public safety, these must be carefully constrained to avoid infringing on the mental and physical integrity of individuals. The ethical justification for any such action must prioritize the preservation of autonomy, consent, and bodily integrity. By emphasizing these principles, society can prevent the normalization of intrusive practices that compromise personal freedom and pave the way for broader encroachments on individual rights. This protective framework ensures that advancements in neuroscience serve humanity without undermining the very principles that uphold human liberty and moral integrity.

Let me finish by clarifying a conceptual difference to show that one can perfectly deny free will while advocating in favor of freedom rights, consent, and autonomy as valuable juridical concepts. It is crucial to distinguish between the metaphysical concept of free will and the juridical concepts of freedom rights, autonomy, and consent. Free will, as traditionally defined, relies on two central principles: the principle of alternative possibilities (the capacity to have chosen otherwise) and ultimate control (being the true in-control originator of one's actions). Under hard incompatibilism, these principles are denied as human behavior is understood to be determined by neurobiological and environmental factors or by randomness, making ultimate control and true alternatives impossible.

However, advocating for freedom rights, autonomy, and consent does not require the existence of metaphysical free will. These rights are valued not because they assume individuals have metaphysical alternatives and ultimate control but because they protect basic natural volition and practical rationality. Even if the world is fundamentally determined—or indetermined—humans still have a basic agency and display a sense of volition and will (even if that will is determined). As argued by Pereboom (2001), Kantian ethics emphasizes that human dignity is based on our capacities for rationality and autonomy. These capacities involve setting ends, choosing means, formulating principles, and committing them to be rationally binding. Importantly, positive freedom—understood as the ability to commit to rational principles—remains intact under hard incompatibilism.

This perspective suggests that, even if our choices are determined by physical laws, the practical capacity to act according to reason and make commitments remains valuable. Therefore, denying free will does not strip individuals of moral dignity or negate the value of autonomy and consent. Freedom rights protect the space for individuals to express their determined will in ways that respect their rational capacities. In this framework, autonomy and consent are meaningful because they safeguard human dignity and promote a system of rights that recognizes individuals as rational agents, capable of making their own decisions within the constraints of a determined—or indetermined—world, while rejecting the possibility of coercive interventions that require a very high "degree of brutality (such as forcibly arresting someone to an operating room to implant a brain stimulation device)" (Borbón, 2024, 3).

6 Conclusion

The analysis presented in this chapter argues that free will skepticism, particularly hard incompatibilism, presents profound challenges to traditional moral and legal frameworks that rely on the assumption of individual moral responsibility. The rejection of a strong metaphysical free will implies that foundational elements of retributive justice, including the concept of culpability, become untenable. Consequently, the justification for punitive measures collapses, urging a re-evaluation of how societies address deviant behavior.

The emergence of neurotechnologies capable of altering cognitive and emotional states, particularly those framed as moral neuroenhancements, adds a layer of complexity to this discussion. While proponents argue for their potential to reduce harm and foster prosocial behavior, this chapter has argued that coercive and nonconsensual neurointerventions threaten fundamental human rights. The ethical dilemma posed by such intervention's centers on autonomy, informed consent, and cognitive liberty. Coercive measures, even when intended for social benefit, risk eroding the principles of human dignity. The essential nucleus of rights such as human dignity, freedom of thought, bodily integrity, and autonomy must be safeguarded as absolute protection to prevent state overreach and ensure that advances in neuroscience do not compromise individual natural freedom.

Neurorights, though proposed as novel protections, may paradoxically undermine existing human rights standards by being framed as overly relative rather than having any component of absolute prohibitions on their essential nucleus. This chapter has shown that new relative rights open the door to potential state misuse and coercive practices under the guise of public safety or utilitarian social benefit. The principle of progressivity in human rights insists that advancements must not erode existing protections but rather enhance them. The current proposals for neurorights, as critiqued herein, may fail to meet this standard by allowing exceptions that weaken essential rights such as freedom of thought and bodily integrity.

In advocating for a more humane, ethical framework, this chapter argues that neuroabolitionism—the gradual replacement of criminal law with systems that emphasize prevention, rehabilitation, and social justice—offers a compelling alternative. This approach aligns with free will skepticism by rejecting retributive punishment and supporting justice systems rooted in dignity, social structural justice, and public health policies. Such a model prioritizes the protection of autonomy, informed consent, and bodily integrity, ensuring that neurotechnological advancements respect and uphold these non-negotiable principles.

In conclusion, the ethical application of neuroscience should pivot away from coercive uses and instead inform a justice system that emphasizes prevention, social justice, well-being, and human dignity. Safeguarding dignity, freedom of thought, bodily integrity, and autonomy while addressing the determinants of antisocial behavior through non-punitive measures represents a future-oriented approach that aligns with both free will skepticism and fundamental human rights. Then, my invitation is to rethink public policies and radically challenge the way in which we currently fail to do justice.

References

Borbón, D. (2022). Neurosociology and penal neuroabolitionism: Rethinking justice with neuroscience. *Frontiers in Sociology, 7*, 814338. https://doi.org/10.3389/fsoc.2022.814338

Borbón, D. (2024). Free will, quarantines, and moral enhancements: Neuroabolitionism as an alternative to criminal law. *Frontiers in Sociology, 9*, 1395986. https://doi.org/10.3389/fsoc.2024.1395986

Brinkley-Rubinstein, L. (2013). Incarceration as a catalyst for worsening health. *Health Justice, 1*(1). https://doi.org/10.1186/2194-7899-1-3

Bublitz, C. (2016). Moral enhancement and mental freedom. *Journal of Applied Philosophy, 33*, 88–106. https://doi.org/10.1111/japp.12108

Bublitz, J. C. (2022). Novel neurorights: From nonsense to substance. *Neuroethics, 15*, 7. https://doi.org/10.1007/s12152-022-09481-3

Bublitz, J. C. (2013). My mind is mine!? Cognitive liberty as a legal concept. In E. Hildt, & A. Franke (Eds.), *Cognitive enhancement. Trends in augmentation of human performance* (Vol. 1). Springer. https://doi.org/10.1007/978-94-007-6253-4_19

Bublitz, J. C. (2021). Freedom of thought as an international human right: Elements of a theory of a living right. In P. Blitz & J. C. Bublitz (Eds.), *The law and ethics of freedom of thought* (Vol. I, pp. 50–101). Palgrave McMillan.

Caruso, G. D. (2021). *Rejecting retributivism: Free will, punishment, and criminal justice.* Cambridge University Press.

Caruso, G. D., & Pereboom, D. (2022). *Moral responsibility reconsidered.* Cambridge University Press. https://doi.org/10.1017/9781009219730

Christie, N. (1981). *Limits to Pain.* Martin Robertson & Company Ltd.

Cid, J. (2009). Is imprisonment criminogenic. *European Journal of Criminology 6*(6), 459–480. https://doi.org/10.1177/1477370809341128
Constantino, P., Assis, S. G., & Pinto, L. W. (2016). The impact of prisons on the mental health of prisoners in the state of Rio de Janeiro. *Brazil. Cien Saude Colet, 21*(7), 2089–2100. https://doi.org/10.1590/1413-81232015217.01222016
Corte Constitucional de Colombia. (1998). *Sentencia T-153 of 1998.*
Corte Constitucional de Colombia. (2013). *Sentencia T-388 of 2013.*
Corte Constitucional de Colombia. (2015). *Sentencia T-762 of 2015.*
Corte Constitucional de Colombia. (2022). *Sentencia SU-122 of 2022.*
Cullen, F. T., Jonson, C. L., & Nagin, D. S. (2011). Prisons do not reduce recidivism. *The Prison Journal, 91* (3 Suppl. l), 48S–65S. https://doi.org/10.1177/0032885511415224
Cunego, A. (2016). *Development policies evaluation under the human rights-based approach.* Berg Institute. https://doi.org/10.2307/j.ctvkwnnrz
Davis, A. (2003). *Are prisons obsolete.* Seven Stories Press.
De Grazia, D. (2014). Moral enhancement, freedom, and what we (should) value in moral behaviour. *Journal of Medical Ethics, 40*, 361–368. https://doi.org/10.1136/medethics-2012-101157
Díaz Soto, J. M., & Borbón, D. (2022). Neurorights vs. neuroprediction and lie detection: The imperative limits to criminal law. *Frontiers in Psychology, 13*, 1030439. https://doi.org/10.3389/fpsyg.2022.1030439
Douglas, T. (2008). Moral enhancement. *Journal of Applied Philosophy, 25*, 228–245. https://doi.org/10.1111/j.1468-5930.2008.00412.x
Douglas, T. (2014). Criminal rehabilitation through medical intervention: Moral liability and the right to bodily integrity. *The Journal of Ethics, 18*(2), 101–122. https://doi.org/10.1007/s10892-014-9161-6
Edgemon, T. G., & Clay-Warner, J. (2019). Inmate mental health and the pains of imprisonment. *Soc. Ment. Health, 9*(1), 33–50. https://doi.org/10.1177/2156869318785424
Fair, H., & Walmsley, R. (2021). *World prison population list.* (13th ed.). Institute for Crime & Justice Policy Research at Birkbeck University of London.
Feldman, D. P. (2002). *Chaos and fractals: An elementary introduction.* Oxford University Press.
Ferrajoli, L. (1995). Derecho y razón: Teoría del garantismo penal. Trotta.
Fins, J. (2022). The unintended consequences of chile's neurorights constitutional reform: Moving beyond negative rights to capabilities. *Neuroethics, 15*, 26. https://doi.org/10.1007/s12152-022-09504-z
Glannon, W. (2011). *Brain, body, and mind: Neuroethics with a human face.* Oxford University Press.
Guggisberg, A. G., Mottaz, A., Spinelli, L., Nahum, L., Seeck, M., & Michel, C. M. (2013). Timing and awareness of movement decisions: Does consciousness really come too late? *Frontiers in Human Neuroscience, 7*, 385. https://doi.org/10.3389/fnhum.2013.00385
Haney, C. (2017). "Madness" and penal confinement: Some observations on mental illness and prison pain. *Punishment & Society, 19*(3), 310–326. https://doi.org/10.1177/1462474517705389
Haney, C. (2012). The psychological effects of imprisonment. In *The Oxford Handbook of Sentencing and Corrections.* Oxford: Oxford University Press. https://doi.org/10.1093/oxfordhb/9780199730148.013.0024
Harris, S. (2012). *Free will.* Free Press.
Herrera-Ferrá, K., Muñoz, J. M., Nicolini, H., Saruwatari Zavala, G., & Martínez Bullé Goyri, V. M. (2022). Contextual and cultural perspectives on neurorights: Reflections toward an international consensus. *AJOB Neurosci*, 1–9. https://doi.org/10.1080/21507740.2022.2048722
Hossenfelder, S., & Palmer, T. (2020). Rethinking superdeterminism. *Frontiers in Physics, 8*, 139. https://doi.org/10.3389/fphy.2020.00139
Huey, M. P., & Mcnulty, T. L. (2005). Institutional conditions and prison suicide: Conditional effects of deprivation and overcrowding. *Prison Journal, 85*(4), 490–514. https://doi.org/10.1177/0032885505282258

IACHR. (1993). *Informe anual: Campos en los cuales han de tomarse medidas para dar mayor vigencia a los derechos humanos, de conformidad con la Declaración Americana de los Derechos y Deberes del Hombrey Convención Americana sobre Derechos Humanos*. IACHR. Available online at: https://www.cidh.oas.org/annualrep/93span/cap.v.htm

IACHR. (2017). Cuadernillo de jurisprudencia de la Corte Interamericana de Derechos Humanos (Nº 9): Personas privadas de libertad.

Ienca, M. (2021). On neurorights. *Frontiers in Human Neuroscience, 15*, 701258. https://doi.org/10.3389/fnhum.2021.701258

Ienca, M., & Andorno, R. (2017). Towards new human rights in the age of neuroscience and neurotechnology. *Life Sci. Soci. Pol., 13*, 5. https://doi.org/10.1186/s40504-017-0050-1

Laplace, P. S. (1951). *A philosophical essay on probabilities (F. W. Truscott & F. L. Emory, Trans.)*. Dover Publications.

Libet, B., Gleason, C. A., Wright, E. W., & Pearl, D. K. (1983). Time of conscious intention to act in relation to onset of cerebral activity (readiness-potential): The unconscious initiation of a freely voluntary act. *Brain, 106*(3), 623–642.

Mathiesen, T. (2006). *Prison on trial*. Waterside Press.

Mathiesen, T. (2015). *The politics of abolition revisited*. Routledge.

Mathiesen, T. (2003). *Juicio a la prisión: una evaluación crítica*. Buenos Aires: EDIAR.

Meijers, J., Harte, J. M., Jonker, F. A., & Meynen, G. (2015). Prison Brain? *Executive Dysfunction in Prisoners. Front. Psychol., 6*, 43. https://doi.org/10.3389/fpsyg.2015.00043

Meijers, J., Harte, J. M., Meynen, G., Cuijpers, P., & Scherder, E. J. A. (2018). Reduced self-control after 3 months of imprisonment; a pilot study. *Frontiers in Psychology, 9*, 69. https://doi.org/10.3389/fpsyg.2018.00069

Mele, A. R. (2014). *Free: Why science hasn't disproved free will*. Oxford University Press.

Moreu Carbonell, E. (2021). The regulation of neuro-rights. *European Review of Digital Administration & Law-Erdal, 2*, 149–162.

National Penitentiary and Prison Institute INPEC. (2025). *Tableros estadísticos—Académico Intramural Nacional*.

Nurse, J., et al. (2003). Influence of Environmental Factors on Mental Health within Prisons: Focus Group Study. *BMJ, 327*(7413), 480. https://doi.org/10.1136/bmj.327.7413.480

Pereboom, D. (2001). *Living without free will*. Cambridge University Press.

Persson, I., & Savulescu, J. (2008). The perils of cognitive enhancement and the urgent imperative to enhance the moral character of humanity. *J. Applied Ethics, 25*, 162–177. https://doi.org/10.1111/j.1468-5930.2008.00410.x

Persson, I., & Savulescu, J. (2012). *Unfit for the future: The need for moral enhancement*. Oxford University Press.

Persson, I., & Savulescu, J. (2013). Should moral bioenhancement be compulsory? Reply to Vojin Rakic. *Journal of Medical Ethics, 40*, 251–252. https://doi.org/10.1136/medethics-2013-101423

Piper, A., & Berle, D. (2019). The association between trauma experienced during incarceration and PTSD outcomes: A systematic review and meta-analysis. *The Journal of Forensic Psychiatry & Psychology, 30*, 854–875. https://doi.org/10.1080/14789949.2019.1639788

Reiter, K., Ventura, J., Lovell, D., Augustine, D., Barragan, M., Blair, T., et al. (2020). Psychological distress in solitary confinement: Symptoms, severity, and prevalence in the United States, 2017–2018. *American Journal of Public Health, 110*(S1), S56–S62. https://doi.org/10.2105/ajph.2019.305375

Ruiz, S., Ramos-Vergara, P., Rodrigo, C., Altermatt, F., Von-Bernhardi, R., Mauricio, C., et al. (2021). Negative effects of the patients' rights law and neuro-rights bill in Chile. *Revista Médica De Chile, 149*, 439–446. https://doi.org/10.4067/s0034-98872021000300439

Schnittker, J., Massoglia, M., & Uggen, C. (2012). Out and down: Incarceration and psychiatric disorders. *Journal of Health and Social Behavior, 53*(4), 448–464. https://doi.org/10.1177/0022146512453928

Sententia, W. (2006). Neuroethical considerations: Cognitive liberty and converging technologies for improving human cognition. *Annals of the New York Academy of Sciences, 1013*(1), 221–228. https://doi.org/10.1196/annals.1305.014

Trevena, J., & Miller, J. (2010). Brain preparation before a voluntary action: Evidence against unconscious movement initiation. *Consciousness and Cognition, 19*(1), 447–456. https://doi.org/10.1016/j.concog.2009.08.006

Tyrer, P., Duggan, C., Cooper, S., Tyrer, H., Swinson, N., & Rutter, D. (2015). The lessons and legacy of the programme for dangerous and severe personality disorders. Personal. *Mental Health, 9*, 96–106. https://doi.org/10.1002/pmh.1293

Wacquant, L. (2009). *Prisons of poverty (Expanded ed.).* University of Minnesota Press.

Yuste, R., Georing, S., Arcas, A., Bi, G., Carmena, J., Carter, A., et al. (2017). Four ethical priorities for neurotechnologies and AI. *Nature, 551*, 159–163. https://doi.org/10.1038/551159a

Neurotechnologies, Human Rights and Criminal Law

Eduardo Demetrio Crespo and Faustino García de la Torre García

Abstract The most recent discoveries in neuroscience have an impact on the basic assumptions of criminal liability as they deal with the subjective aspects of human behaviour. The neurosciences are coming to new and revealing conclusions thanks to the development of "neurotechnologies," which may prove to be of utility to the justice system but could also serve to commit crimes. This contribution sets out the main challenges for Criminal Law systems in the wake of advances in the neurosciences and the development of technologies applied to the study of the human brain. Some relevant questions about the epistemological and axiological status of criminal law in the era of neurotechnologies are addressed.

Keywords Neurorights · Neurotechnologies · Human rights · Criminal responsibility · Neurocrimes

1 Introduction

Advances in neuroscience are especially important to law in general, as its object of study focuses on human behaviour, which legal systems seek to regulate (Ienca & Andorno, 2021). These findings on human behaviour have important implications for criminal law, whose legal consequences are imposed based on subjective aspects of (criminal) human behaviour. The rapid progress in the neurosciences in the last two

This publication is part of the Research Project 'Neuro-Human Rights and Criminal Law' (Neuro-Derechos Humanos y Derecho penal) (PID2023-149978NB-I00), funded by MICIU/AEI/ 10.13039/501100011033 and by ERDF/EU.

E. Demetrio Crespo (✉)
Faculty of Legal and Social Sciences, University of Castilla-La Mancha, Toledo, Spain
e-mail: eduardo.Demetrio@uclm.es

F. G. de la Torre García
Universidad de Castilla-La Mancha, Toledo, Spain
e-mail: Faustino.GarciaTorre@uclm.es

P. López-Silva (ed.), *Contextualizing Neuroprotection: Latin American Perspectives on the Impact of Neurotechnological Development in Life and Society*, Ethics of Science and Technology Assessment 54, https://doi.org/10.1007/978-3-031-96055-0_3

decades and the conclusions that have been drawn from it about human behaviour raise numerous questions that pose a significant impact on the ascription of guilt and the conception of criminal action, the content of which must be discussed considering neuroscientific discoveries, without, in any case, transcending the framework of philosophical reflections on the human person.

The development of technologies applied to the study of the human brain and mind must take this into account. Developments in this field open another range of concerns related to the misuses of neurotechnology, which are already being addressed by lawmakers in different countries as well as by international and regional bodies. This debate also affects criminal law systems, not only in terms of their model of judicial procedures, but also in relation to the ways in which crimes are committed. Given the close relationship between Criminal Law and Human Rights, the new proposals for 'neuro-specific' rights force us to rethink the legal guarantees of citizens in the face of the threats associated with neurotechnologies in the hands of authorities and third parties.

The following pages set forth some general postulations of epistemological challenges related to aspects like criminal action, guilt, subjectivity and identity, which permeate this problem (Demetrio Crespo, 2013a, 2013b; Demetrio Crespo, 2017). The basic axiological issues that the proclamation of neuro-human rights has for Criminal Law are also addressed, as rights of defense against the State and as duties to protect, such as the dangers posed using neurotechnologies by criminal justice systems and by third parties for criminal purposes.

2 The Epistemological Aspects of Criminal Law in the Age of Neurotechnologies

2.1 Action, Willfulness, Responsibility

The so-called 'Theory of crime' (*Verbrechenslehre*) is the area in which the technical changes that can affect the application of criminal law are developed. It constitutes a valuable "toolbox" in which each piece must have a very precise and previously defined use (Demetrio Crespo, 2022, 17 et seq.). The recent question is whether scientific developments—in our case, those coming from brain research—can contribute to a better delineation of the basic categories of the theory of crime. The debate in recent years has largely focused on the field of culpability (Demetrio Crespo, 2013a, 2013b), but other previous elements (Detlefsen, 2006), such as the legal/penal concept of voluntary action itself, are also directly affected, such that, in reality, it is necessary to refer to human behavior in its entirety (Demetrio Crespo, 2022).

There exists a particularly difficult problem with severe practical consequences when it comes to investigating the delimitation between the incapacity for action and culpability (Alonso Alamo 2022, 329 et seq). Key in this field is the interaction, at the philosophical-conceptual level, between the terms "action," "knowledge" and "will" (Hyman, 2015), as well as between "motives," "reasons" and "causes" (Habermas, 2004; Patterson, 2019) and, in particular, between "motives" and "emotions" (Peralta, 2012; Alonso 2016, 26 ff.), within the multiplicity of subjective elements that can play a role in the determination of criminal responsibility, with respect to which a psychological basis (increasingly accessible) is combined with the normative consideration of them (Díez 1990, 323 et seq.; Kelker 2007, 521 et seq.).

In any case, there is no doubt that human action cannot be understood—as the most "physicalist" positions of neuroscience would assert—only in a causal way. Rather, it must be interpreted in the complexity of its context and, as Alonso (Alonso Álamo, 2022) emphasizes, in terms of the content of the will. The challenges for the theory of criminal action go further today due to the rise of artificial intelligence, which tests the capacity of the current system of criminal law attribution. Thus, it is worth asking, for example, how the penal system should react when it comes to undervalued results of notable importance that were not considered by programmers and, in turn, whether this reality allows us to speak of a responsibility gap (Busato, 2022; Hallevy, 2013).

2.2 *Subjectivity, Subjective Responsibility and the Subject of the Action*

It could be said that, in a way, the main concern of the developments and proposals emanating from the experimental and technological field is that they call into question an elementary principle of democratic criminal law: subjective responsibility. The punitive model of the constitutional and democratic State does not prohibit the production of results, as such, unless they can be subjectively linked to their perpetrator, for which it is necessary for him or her to have produced them—by action or omission—intentionally or negligently.

The significant disparity in penalties prescribed by criminal codes for intentional versus negligent behavior already establishes a decisive difference that must be clearly defined based on the degree of accountability of the individual for their actions. This distinction cannot be overlooked in any case. However, traditional criminal law theory is somewhat ineffective in specifying the criteria that differentiate these two categories. Therefore, it is relevant to explore whether cognitive neuroscience can provide some support in this regard.

In accordance with the astute reflections of Pérez Manzano (2022, 395), as in general the criminalist cannot be ignorant of scientific advances, neither should he do so with respect to those branches of knowledge that have as their object the study of mental states and cognitive processes when it comes to configuring a concept as fundamental to determining criminal responsibility as willful misconduct. The latter implies, of course, a prior position, which we share, on the dual nature of the latter, which in no case can be interpreted as a purely normative construct, but rather, like all other concepts of the theory of crime, must refer to a reality of a factual nature capable of being proven in criminal proceedings (Paredes Castañón, 2020, 875 et seq.; Pérez Manzano, 2021, 553 et seq).

2.3 The Epistemic and Technical/legal Debate

Criminal action is contemplated, in turn, from a prospective perspective insofar as it analyzes the epistemic value of forecasts after the emergence of algorithmic tools in criminal justice, which takes on an epistemological nature not seen before. Indeed, there are many inference judgments intimately related to the search for cause-effect relationships that legal operators are obliged to make without being obliged to follow any scientific method previously determined for this purpose. In the coming years decisions will have to made as regards the exhilarating question, posed by Miró and Castro Toledo (2022) of whether the penal system can rely, for example, on supercomputing strategies, based on Machine Learning, to make decisions. As they rightly point out, traditional "dogma" cannot be limited to showing the limitations of existing techniques. Rather, it is faced with the difficult task of trying to provide reliable information to those who must decide specific cases so that guarantees are respected to the furthest extent possible.

As we have been pointing out, the challenges that new neurotechnological and algorithmic advances entail for the Criminal Law systems will force a whole series of deep reflections on its progressive incorporation, without losing sight, in any case, of the Human Rights that may be affected. One of the most promising and, at the same time, problematic areas is that of Predictive Sentencing (Keijser et al., 2019; Ryberg, 2020). Although opening new perspectives that may prove useful to judicial decision-making, at this time both the scope and the reliability of the potentially applicable mechanisms, with respect to traditional methods, are very uncertain (Grosse-Wilde, 2022; Martínez Garay, 2020). At the same time, there are voices warning about the dangers of the model of Machine Learning algorithms for criminal justice, due to a lack of transparency and reproducibility, among other things, they do not have the capacity to explain the results achieved, such that they could become genuine "black boxes" (Martínez Garay, 2022).

From a procedural point of view, it is a question of whether the incorporation of Artificial Intelligence can contribute to making the penal system work in a less arbitrary way than it does today (Nieva Fenoll, 2018). Pastor (2022, 531-548), while noting that the most interesting experiment does not really consist of asking whether it is possible to administer justice the way it always has been, but better, but rather of trying to do it in a completely different way from everything we know, striving to emphasize that, in any case, the goals to be pursued must remain identical. These are none other than those of the constitutional rule of law to seek unrestricted respect for Human Rights.

2.4 Neurological Identity as a Fundamental Core

2.4.1 Conscious Acts and the Identity of the Human Being

It remains, however, to ascertain whether this whole problem affects the identity of the human being himself, and to what extent Neuroscience questions or endangers it (Demetrio Crespo, 2013a, 2013b, 237 et seq.). It is not the same for neuroscience to question human identity, nor does it have the same scope as for it to *endanger* it. From the first point of view, the discussion corresponds, in general, with the nature of conscious acts and the identity of the human being as an individual, but also as a social being. Another very different question is whether it endangers it. If the possibilities of intervening in the brain through the various techniques available (from 'neuroimaging' to 'neurosurgery' to neuropharmaceuticals and others, as discussed in the field of 'neuroenhancement') make it possible to 'manipulate' neurological identity, the problem becomes how to protect what makes us unique by virtue of our personality.

The questioning of the identity of the subject, understood from the point of view of the conformation of his personality or individuality as a human being, would come by way of assuming a purely physical neurodeterminism in which the desires, motives and intentions (which according to popular psychology, underpin conscious decisions) do not have causal efficacy, but rather decisions are, so to speak, pre-programmed in the limbic system (Morse, 2011, 118–120). It is therefore necessary to ask whether this image of the human person really corresponds to the biological configuration or design of the brain, which, with the sole knowledge of a lawyer, is impossible. However, it is pertinent, and even necessary, to debate whether the data we have to date allow us to speak of conscious or intentional decisions. The answer at this point, as Pérez Manzano explains, with a profusion of arguments (2013, 120 et seq.), is clearly negative.

As for what it identifies—that is, that which confers an identity, in principle, unique—each human being, we could say that there is not at this time moment a convincing neurobiological explanation of how what we might call the feeling of "self" or the feeling of "selfhood" arises. Particularly clear in this field is the neuropsychologist Wolfgang Prinz, who believes that the performance of the brain can be analyzed from three perspectives: behavior, representation and subjectivity. While quite a bit is already known about the first two aspects, although the basic principles are still far from understood, science feels powerless in the latter field. This is because it cannot explain how the brain processes that produce consciousness work. In the opinion of the author, the nature of subjectivity and consciousness cannot be explained and understood only with the "arsenal" of research on the brain. Rather, a complete framework theory is needed that understands and considers not only its biological and natural foundations, but also the social and cultural ones on which both are based (Prinz, 2004, 35).

2.4.2 Intersubjectivity, Intention and Criminal Liability

It is precisely the intersubjective elements that have an important role whose true scope is not explained by the experiments of Libet (Demetrio Crespo, 2011). In the opinion of Pérez Manzano, the most relevant objections (2013, 123 et seq.) to the latter refer to the extrapolation of the results obtained from the observation of very simple actions to much more complex areas of human action, and, more specifically, to their limited usefulness in explaining human action in normative contexts.

In this regard, Habermas (2004, 872) points out that the objectifying language of Neurobiology demands of the brain the grammatical role that the "I" had previously played, but without finding any correspondence in the language of everyday psychology. Given this, the philosopher asks the following question: is the deterministic conception really a thesis based on the natural scientific method, or is it only part of a naturalistic worldview arising from a speculative interpretation of scientific knowledge? Whereas Kant would have tried to reconcile the causality that comes from freedom with natural causality at the price of a dualism between the worlds of the intelligible and of phenomena, reductionist research programs could only circumvent the difficulty of a dualism of explanatory perspectives and language games at the price of epiphenomenalism. This is precisely due to a fundamental reason, which is that "normally actions are the result of a complex concatenation of intentions and thoughts, which weigh alternative ends and means in the light of opportunities, resources and obstacles" (Habermas, 2004, 873).

A question, perhaps less abstract and with a clear legal impact, is to what extent advances in Neuroscience can be considered in the Law and, more importantly, how far they can go. In this respect, Winfried Hassemer made it clear long ago that we should not be allowed to do everything we *could* do, which is a cornerstone of the regulation of the limits of knowledge in criminal proceedings. In this way, the author tries to argue that the search for truth in criminal proceedings does indeed take place under stable and well-founded conditions, which necessarily prevent things from being clarified as comprehensively, completely and as close to reality as the natural sciences are accustomed to.

Demonstrating this are the numerous rights of the defense established by the criminal process itself, such as the right not to plead guilty, or to remain silent, based on three principles: the legitimate interests of the people affected by the investigation, the general interest in the existence of certain professions and institutions, and the principle of due process (Hassemer, 2009, p. 829 et seq; Demetrio Crespo, 2011).

3 Axiological Aspects of Criminal Law in the Era of Neurotechnology

3.1 Neurorights. Specific Human Rights?

After years of skepticism, Neurorights are beginning to be taken "seriously". The impetus of several groups of interdisciplinary specialists, who have been expressing the need to guarantee, by law, the properties of the human brain and mind, has been decisive in persuading various international, regional and national organizations to begin to confront the emerging advance of neurotechnology in various areas of social relations. As a result of the ethical reflections prompted by the development and application of neurotechnologies, a new generation of Human Rights is emerging to deal with the unique and unprecedented problems posed by access to the brain activity of humans. Although "Neurorights" are, for the time being, a series of directives that are not binding on States and private agents, there is no doubt that their guidelines will inspire the values on which the legislative frameworks of impending regulations will be developed (Yuste, 2023, 2871; Reche Tello, 2024, 71 et seq).

In the last two decades there have been the greatest advances in history in research on the human brain, through the development of a series of technologies that are able to access, record and influence neural processes (Herz, 2023; Wajnerman & López-Silva, 2022). The improvement of these techniques with Artificial Intelligence methods will serve, in medicine, to effectively treat common ailments such as migraines, or more complex diseases, such as Parkinson's; to deal with strokes more quickly, and reliably, and even to tackle brain cancer. Prosthesis users will be able to remotely control these devices, and the public will be replacing control devices with brain-computer interfaces to experience more freedom and independence than ever before (Muñoz et al., 2023, 2).

However, specialists warn that neurotechnologies are a "neutral" tool (Yuste, 2019, 48; 2023, 2870; Yuste & Salcedo 2023, 16) that could also be used to do the kind of damage that we have only associated with science fiction literature (Llamas & Marinaro, 2021, 91; Gulyaeva & Farinella, 2022, 283-284). Numerous studies show that the potential of these techniques could turn brainwashing and mind control practices into reality. They could be used for interrogations, or for discrimination, to manipulate the minds of people, to provoke emotions, guide preferences, or manipulate memory and will, among other cognitive-emotional phenomena. The possibility of accessing, recording and influencing the human brain and mind forces lawmakers to determine the boundaries at which it is legitimate to try to mitigate the unintended consequences of neurotechnologies and to prevent malicious uses of these techniques. According to the conclusions reached in various interdisciplinary forums at which the ethical issues related to the development and application of neurotechnologies have been discussed, it is necessary and urgent to establish legal protection for privacy, integrity and mental self-determination in the face of the new challenges of the neurotechnological era (Bublitz & Reinhard, 2014; Ienca & Andorno, 2021; Ienca, 2021a, 2021b; Yuste et al., 2021).

What have been called "neurorights" are a "battle cry" that has emerged in the last decade in favor of legally protecting the domain of the human brain and mind from the dangers posed by the development and application of neurotechnologies (Ienca, 2021a, 2021b; Yuste et al., 2021). First, Neurorights advocates aim to update Human Rights catalogues to include the benefits and properties of the human brain and mind, to head into the future with a robust protection of human nature (Brown, 2024; Herz, 2023). Within these stances, two positions can be identified: a more "innovative" or "positive" vision, which advances the need to create new rights; and a more "conservative" or "negative" one, which affirms the possibility that existing rights can cover the protective needs of the human brain and mind (Kellmeyer, 2022, 421; Ligthart et al., 2023, 464–465; Brown, 2024, 3).

Neurorights have, so far, been considered ethical, social or natural principles that States, and third parties cannot be compelled to observe and respect (Ienca, 2021a, 2021b). However, concerns about the ethical challenges posed by the use of neurotechnologies have grown, as advances have been made in the development of these techniques, which has led to the emergence of a series of regulations at different levels of the legal system, and with varying degrees of enforceability, which, together, form the values, principles and directives for action that will guide the development of "neurolaw" (García-López, 2022) in national regulations, including Criminal Law.

3.2 *Neurorights and Criminal Law*

The punitive systems of States will not remain impassive in the face of the recognition, in one way or another, of Neurorights, since the relationship between Human Rights and Criminal Law is especially close. The Criminal Law systems have historically served to deny individuals the enjoyment of freedoms. Human rights have thus been conceived as negative obligations of states, requiring authorities to refrain from certain actions. But individuals, or third parties, have also been denying Human Rights, which are also conceived as positive State obligations, to do or not abstain from certain actions. Neuro human rights will have a significant impact on the foundations and limits of Criminal Law.

There is no doubt that authorities (States) represent the main threat to the enjoyment of Neurorights, because authorities have a monopoly on legitimate force, without prejudice to the fact that each coercive resource requires a justification. The fact that the main world powers are developing large projects in relation to the study of the human brain (such as the Brain initiative in the United States; the Human Brain project in the European Union; the China Brain project in China; the Japan Brain project in Japan; the Korean Brain initiative in South Korea; the Brain Research strategy in Canada; and the Australian Brain alliance in Australia) whose conclusions could completely transform the application of criminal justice as we know it, should alert us to the current content of "defense rights" (*Abwehrrechten*), which protect the human brain and mind against Criminal Law.

However, in the last fifty years powerful competitors of the Leviathan have emerged in terms of forces threatening freedoms: terrorist or criminal organizations, multinational companies and other third non-state actors that also can seriously endanger legal assets related to Neurorights. As large multinational companies—such as Neuralink or Meta—are developing neurotechnologies for the public in aspects of daily life, which can create huge gaps for the security of the individuals or people, defenders of Neurorights argue that legal assets related to the brain and the human mind must also be protected against attacks by non-state actors. As the main guardians of security and guarantors of peace, nation states cannot remain passive in the face of malicious uses of neurotechnologies by individuals, so it is also appropriate to address possible "State duties to protect" (*grundrechtliche Schutzpflichten*) contained in their proposals, with a view to the protection of the human brain and mind by means of Criminal Law.

3.3 The Defense of the Human Brain and Mind Against Neurotechnologies Through Criminal Law

3.3.1 The Use of Neurotechnologies for the Prevention and Suppression of Crime

One of the social areas in which it is estimated that neurotechnologies will trigger a real revolution is criminal justice, in every one of its phases. As a preventive measure, the possibility of developing techniques aimed at crime prevention based on studies related to the human brain and mind, capable of determining which people may commit crimes, or be repeat offenders, has been proposed for years. Neurotechnologies have a high potential to be used as lie detectors, diagnose brain abnormalities, predict future dangerousness, determine guilt or penalties, or generally assess criminal responsibility. Finally, these techniques could be used to decide on the most appropriate prison treatments for a given person (Llamas & Marinaro, 2021, 101 et seq).

Unlike the use of neurotechnologies in other areas of social life, such as medicine, leisure or sports, the use of these techniques in criminal justice could be applied in a coercive way seriously endangering the rights of defense of the accused, unacceptable in a democratic state of constitutional law (Lingthart, 2022). Brain reading in criminal justice could be coercive, such as when physical coercion, including mechanical or chemical immobilization, is used. Coercive brain reading could cause physical and/or mental harm, or be physically intrusive, involving exposure to radiation, or entailing health risks, which would violate the right to physical or mental wellness and prohibitions of torture and degrading treatment. These methods could also interfere with the right to mental privacy. In Criminal Law, the use of coercive techniques for interrogation, forcing the subject to express himself, or to transmit, impart or disseminate information, ideas or opinions, through a coercive reading of the brain, may violate the rights of the accused, without prejudice to their justification. This could even be achieved without the need to make the subject speak, but by "reading" his or her mind.

Brain-based lie detection has been applied using fMRI, PET, SPECT, and EEG. CIT is also put forth to prove knowledge of guilt. This test makes it possible to detect the recognition of specific information, such as the central aspects of a crime. In addition to investigating past crimes, CIT could also help identify knowledge of a future crime. The results of different neuroimaging technologies, such as Magnetic Resonance Imaging (MRIs), positron emission tomography, and electroencephalograms, have already been used to evaluate legal dementia in specific criminal cases (Lingthart, 2022, 53, 68, 99, 110). Thus, neuro human rights should prohibit the imposition of sanctions or other negative consequences for holding certain thoughts or statements or being in a specific state of mind (Herz, 2023, 7).

3.3.2 The Use of Neurotechnologies for the Commission of Crimes

While neurotechnologies discover the neural functions and processes on which we can act, so that people enjoy more freedom and independence than ever before, the "weak points" of the brain through which the human mind can be deceived and manipulated are being revealed. Implantable neurotechnologies that are currently applied to deal with neurodegenerative diseases, to control electronic devices in smart homes, vehicles, drones and prostheses, open the door to attacks that seriously endanger the safety of customers (Pazouki et al., 2022; Armengol-Urpi et al., 2023). It is possible, by means of these techniques, to extract and inject sensitive information into the brain of a person to manipulate their behavior (Pugh et al., 2018, 221); or cause them pain, or impulse control disorders, or alterations of their emotional effects (Fernández-García et al., 2023, 120).

The notion of neurological implants being used against their users no longer belongs to the world of fantasy. Several studies assert that the increasing sophistication of techniques such as "invasive neuromodulation" has created a small but real risk of unauthorized access to implantable pulse generators, opening the possibility of doing serious damage, which invites ethical and legal reflection on the management of neurotechnologies (Pycroft et al. 2016, 454; Pugh et al., 2018). For a few years now, it has been estimated that "neurocrime" is a reality, which refers to "targeted and intentional attacks against neuronal information, taking advantage of the vulnerability of neurotechnologies (Ienca, 2015)". More recently, neurocrime has been defined as actions that directly or indirectly produce harmful or dangerous consequences for the safety or privacy of individuals. Apparently, for now, most of what is called neurocrime can only be perpetrated through cyberattacks (Castro, 2024, 48 et seq).

In this regard, neuro-cybersecurity experts use the term "brain hacking" to refer to unauthorized access to neural devices implanted or used in another person to obtain their neurodata, or to control or manipulate some of their brain functions through stimulation. Such activity may harm or seriously endanger the legal interests of privacy, intimacy and mental autonomy of individuals. Attacks can access sensitive information stored on the device, such as the patient's name, diagnosis, stimulation parameters, and their doctor's data, as well as cause them physical and psychological harm. Hackers could exert a direct influence on an individual's deliberations through brainjacking, exercising direct control over the same processes that underpin our status as autonomous decision-makers (Pugh et al., 2018, 221 et seq).

Some studies have documented successful attacks on the security of various neurotechnologies, such as a virtual keyboard speller, a drone control interface, and a neurofeedback meditation interface, being able to successfully take control of BCIs by typing any phrase into the speller, crashing the drone, and inducing a signal indicating a false meditative state, respectively (Armengol-Urpi et al., 2023). Attacks on BCI and EPG neurotechnologies have also been documented that affect communications between the brain and the system, depleting the battery and draining the brain implants (Pazouki et al., 2022). Other researchers have successfully tested the feasibility of presenting malicious visual stimuli to extract sensitive data from subjects, such as thoughts in BCI technologies (López Bernal et al., 2020). Sensitive information has even been extracted from users, such as credit card passwords, presenting visual stimuli and analyzing their potential response (Armengol-Urpi et al., 2023). All these studies show that these Deep Learning models can be easily "cheated".

The specialists assert that there is a variety and complexity of "neuromodulation" attacks, with their methods increasing in number. According to them, the simplest are called "blind attacks", because the attacker does not require specific knowledge of the patient. It is enough for them to disconnect the stimulation to lose the therapeutic effect. If temporary, these attacks can cause discomfort, as the patient may restart stimulation, although the sudden cessation of the stimulation could cause severe "rebound effect" symptoms with various disorders, such as Parkinson's disease, essential tremor, and obsessive-compulsive disorder. If the effects of the attack are enduring, however, it would be possible to induce tissue damage because of the increase in pulse width and voltage. Blind attacks on neurotechnologies also allow information to be obtained from the implant by passively listening to information transmitted during its normal operation, or by actively accessing the device to receive information (Pycroft et al. 2016, 455).

The most elaborate attacks, which require a certain degree of pathophysiological knowledge of the patient, are called "target attacks". These make use of the devices implanted in a person to alter their behavior and cognition, modifying the stimulation parameters. These attacks can diminish the effect of treatment or induce unpleasant off-target effects by stimulating surrounding structures, as well as obtaining medical records through medical database breaches, social engineering, or simple attacks. Impairments of motor function can also be caused, it being possible to substantially damage motor function by altering basic stimulation parameters, thus increasing the patient's parkinsonian symptoms beyond baseline levels, or aggressors may significantly exacerbate the symptoms of tremor, or induce pain by altering the frequency of stimulation. The deliberate stimulation of inappropriate electrode contacts by an attacker may, therefore, result in personally and socially undesirable emotional changes, which would likely be very distressing to a patient and their loved ones (Pycroft et al., 2016, 456 et seq).

4 Conclusions

The current discussion of Neurorights should be situated in a context that adequately explains where we come from. This issue is important to prevent us from losing sight of previous problematics. Thus, at least two perspectives are relevant to the Criminal Law approach: one that refers to the traditional relationship between the neurosciences and criminal law, and one that has to do with the broader intersection that results, in turn, from the connection with Artificial Intelligence and its use for the generation of neurotechnologies. In addition, but not less important, it is also important to refer to the tense and far-from-obvious correlation between Human Rights and Criminal Law.

This is the task that we have undertaken in an introductory manner in this article, starting by referring to neurotechnologies and Criminal Law in the technical field of crime theory. From this perspective, it is possible to catalogue two conceptual triads that have been questioned by brain research, although without ever yielding the meaning they have always had when assigning criminal responsibility. These are none other than those referring to "action, willfulness and responsibility", on the one hand, and to "subjectivity, subjective responsibility and subject of action", on the other.

Upon examination, the debate that is now being waged in the conceptual framework of "Neurorights" is closely related to the one that has been raging in scientific discussions over the previous two decades. To explain, the breakneck speed of technological development, and the most acute fears about the dangers that research on the brain entails have not yielded any substantial change in the anthropological conception of the human being, with all its ramifications, but rather have transmuted into a danger threatening the human being, as a biophysical entity, and his most fundamental rights.

More than a decade ago we intuited that it is one thing for certain currents of neuroscience to question the identity of the human being himself by "reducing" him to a series of correlations of his own neurological architecture, and another thing for them to be in a technical position to endanger it. That moment has come and, therefore, it is now a question of how to protect what makes us unique by virtue of our personality, so it is no coincidence that we have placed the right to our own neurological identity at the forefront as an essential nucleus.

However, as important as this dimension is that of the always-controversial relationship between Human Rights and Criminal Law, now entering a new stage revolving around the reconceptualization of Human Rights that have already been enshrined and/or, as already happened in the field of genetics, the creation of "neurospecific human rights". This has also been contextualized by referring to the broad field, currently under development, of "neuroprediction, the protection of Fundamental Rights and the penal system," which is related, in turn, to the scientific method that the justice system currently uses to make decisions directly affecting people's rights.

It is clear, in this field, that neurotechnologies have advanced enough to endanger them, and it is for this reason that this work points to the urgent need to consider Neurorights beyond a purely ethical dimension, as nation states cannot remain passive, but rather must be proactive in defending these rights, in a dual dimension. Firstly, the defense against criminal law; that is, against the coercive use of neurotechnologies by the State through their introduction into the justice system with the aim of preventing crime; and, secondly, against their use by third parties for criminal purposes. In this regard, doctrinal and legislative proposals have begun to be proposed on potential "neurocrimes" that, for reasons of space, we have not been able to address here

References

Alonso Alamo, M. (2016). Culpabilidad jurídico-penal, motivos y emociones. La existencia enigmática de la libertad de voluntad. *Revista Penal, 38*, 5–39.

Alonso Alamo, M. (2022). Acción y derecho penal: Problemas de delimitación entre capacidad de acción y capacidad de culpabilidad. In E. Demetrio Crespo (Ed.), *Derecho penal y Comportamiento Humano. Avances desde la Neurociencia y la Inteligencia Artificial* (pp. 329–347). Tirant lo Blanch.

Armengol-Urpi, A., Kovacs, R., & Sarma, S. (2023). Brain-hack: Remotely injecting false brainwaves with RF to take control of a brain-computer interface. *CPSIoTSec*, 53–66.

Brown, C. M. L. (2024). Neurorights, mental privacy and mind reading. *Neuroethics, 34*, 1–19.

Bublitz, J. C., & Reinhard, M. (2014). Crimes against minds: On mental manipulations. *Crim Law and Philos, 8*, 51–77.

Busato, P. C. (2022). De máquinas y seres vivos: ¿Quién actúa en los resultados delictivos derivados de decisiones cibernéticas? In E. Demetrio Crespo (Ed.), *Derecho penal y comportamiento humano. Avances desde la neurociencia y la Inteligencia artificial* (pp. 349–376). Tirant lo Blanch.

Castro, P. A. (2024). What neurohacking can tell us about the mind: Cybercrime, mind upload and the artificial extended mind. In *Challenges of the technological mind. Between philosophy and technology* (pp. 43–62). Springer.

Demetrio Crespo, E (dir). (2013). *Neurociencias y Derecho penal. Nuevas perspectivas en el ámbito de la culpabilidad y tratamiento jurídico-penal de la peligrosidad*. Edisofer/BdeF, Madrid/ Buenos Aires.

Demetrio Crespo, E (dir). (2022). *Derecho penal y Comportamiento Humano. Avances desde la Neurociencia y la Inteligencia Artificial*. Tirant lo Blanch.

Demetrio Crespo, E. (2011). Libertad de voluntad, investigación sobre el cerebro y responsabilidad penal. *InDret, 2*, 1–38.

Demetrio Crespo, E. (2013). Compatibilismo humanista. Una propuesta de conciliación entre Neurociencias y Derecho penal. In Demetrio Crespo, E (Ed.), *Neurociencias y Derecho penal. Nuevas perspectivas en el ámbito de la culpabilidad y tratamiento jurídico-penal de la peligrosidad* (pp. 17–42). Edisofer/BdeF, Montevideo-Buenos Aires/Madrid.

Demetrio Crespo, E. (2013). Identidad y responsabilidad penal. *Anuario de la Facultad de Derecho de la Universidad Autónoma de Madrid, 17*, 237–254.

Demetrio Crespo, E. (2017). *Fragmentos sobre Neurociencias y Derecho Penal*. BdeF, Montevideo-Buenos Aires.
Demetrio Crespo, E. (2022). El Derecho penal ante el desafío neurotecnológico y algorítmico. In Demetrio Crespo, E (Ed.), *Derecho penal y Comportamiento Humano. Avances desde la Neurociencia y la Inteligencia Artificial* (pp. 17–35). Tirant lo Blanch.
Detlefsen, G. (2006). *Grenzen der Freiheit - Bedingungen des Handelns—Perspektive des Schuldprinzips*. Duncker & Humblot.
Díez Ripollés, J. L. (1990). *Los elementos subjetivos del delito: Bases metodológicas*. Tirant lo Blanch, Valencia.
Fernández-García, C. E., Vargas Marín, D., & Sotomayor Velásquez, H. (2023). Neurohacking and artificial intelligence in the vulnerability of the human brain: Are we facing a threat?. In *Artificial intelligence in higher education and scientific research. Future Development* (pp. 117–128). Springer.
García-López, E. (2022). Neuroderecho y neuroderechos en Hispanoamérica. In E. Demetrio Crespo (Ed.), *Derecho penal y Comportamiento Humano. Avances desde la neurociencia y la inteligencia artificial* (pp. 713–732). Tirant lo Blanch.
Grosse-Wilde, T .(2022). "Predictive Sentencing"—Just a hype or a revolution? In E. Demetrio Crespo (Ed.), *Derecho penal y Comportamiento Humano. Avances desde la Neurociencia y la Inteligencia Artificial*. Tirant lo Blanch.
Gulyaeva, E. E., Farinella, F. (2022). Human neuro-rights. *Rev Quaestio Iuris, 15*(1), 278–299.
Habermas, J. (2004). Freiheit und determinismus. *Deutsche Zeitschrift Für Philosophie, 52*(6), 871–890.
Hallevy, G. (2013). *When robots kill. Artificial intelligence under criminal law*. Northeastern University Press, Boston.
Hassemer, W. (2009). Grenzen des Wissens im Strafprozess. *ZStW, 121*, 829–859.
Herz, N. (2023). Neurorights. Do we need new human rights? A reconsideration of the right of thought. *Neuroethics, 16*, 1–15.
Hyman, J. (2015). *Action, knowledge & will*. Oxford University Press, Oxford.
Ienca, M. (2015). Neuroprivacy, neurosecurity and brain-hacking: Emerging issues in neural engineering. *Bioethica Forum, 8*(2), 51–53.
Ienca, M. (2021a). Neuroderechos: ¿por qué debemos actuar antes de que sea demasiado tarde? *Anuario Internacional CIDOB, 1*, 42–43.
Ienca, M. (2021b). On Neurorights. *Frontiers, 24*, 1–11.
Ienca, M., & Andorno, R. (2021). Hacia nuevos Derechos Humanos en la era de la Neurociencia y la Neurotecnología. *Análisis Filosófico, 41*(1), 141–185.
Keijser, J. W., Roberts, J. V., & Ryberg, J. (2019). *Predictive sentencing. Normative and empirical perspectives*. Hart Publishing, Oxford.
Kelker, B. (2007). *Zur Legitimität von Gesinnungsmerkmalen im Strafrecht*. Vittorio Klostermann, Frankfurt am Main.
Kellmeyer, P. (2022). A human rights-based approach for governing neurotechnologies. *The Cambridge handbook of responsible artificial intelligence* (pp. 412–426). Cambridge University Press.
Lingthart, S. (2022). *Coercive brain-reading in criminal justice. An analysis of european human rights law*. Cambridge University Press, Cambridge.
Ligthart, S., et al. (2023). Minding rights: Mapping ethical and legal foundations of 'Neurorights.' *Cambridge Quarterly of Healthcare Ethics, 32*(4), 461–481.
Llamas, N. E., & Marinaro, J. Á. (2021). Neuroderecho: Adaptabilidad de la normativa de derechos humanos con relación a las nuevas neurotecnologías y propuestas para su ampliación. *SCIO. Revista De Filosofía, 21*, 83–111.
López Bernal, S., et al. (2020). Cyberattacks on miniature brain implants to disrupt spontaneous neural signaling. *IEEE Access, 8*, 152204–152221.

Martínez Garay, L. (2020). Evidence-based sentencing y evidencia científica. A la vez, algunas consideraciones sobre las «políticas basadas en la evidencia» y el Derecho Penal. *Teoría y Derecho. Revista de Pensamiento Jurídico 28*, 238–276.

Martínez Garay, L. (2022). ¿Ciencia o alquimia? Algoritmos y transparencia en la valoración del riesgo de reincidencia. In *Derecho penal y Comportamiento Humano. Avances desde la Neurociencia y la Inteligencia Artificial* (pp. 489–506). Tirant lo Blanch.

Miró Llinares, F., Castro Toledo, F. J. (2022). ¿Correlación no implica causalidad? El valor de las predicciones algorítmicas en el sistema penal a propósito del debate epistemológico sobre el "fin de la teoría". In *Derecho penal y Comportamiento Humano. Avances desde la Neurociencia y la Inteligencia Artificial*. Tirant lo Blanch.

Morse, S. (2011). Neuroscience and the future of personhood and responsibility. In *Constitution 3.0: Freedom and technological change* (pp. 113–129). Brookings Institution Press.

Muñoz, J. M., Bernácer, J., & Güell, F. (2023). A conceptual framework to safeguard the neuroright to personal autonomy. *Neuroethics, 16*(18), 1–13.

Nieva Fenoll, J. (2018). *Inteligencia artificial y proceso judicial*. Marcial Pons, Madrid-Barcelona-Buenos Aires-São Paulo.

Paredes Castañon, J. M. (2001). Problemas metodológicos en la prueba del dolo. *Anuario De Filosofía Del Derecho, 18*, 67–94.

Paredes Castañón, J. M. (2020). Dolo y psicología de sentido común. *Libro Homenaje al Profesor Diego-Manuel Luzón Peña con motivo de su 70º aniversario* (pp. 875–885). Reus.

Pastor, D. (2022). ¿Sueña el sistema penal con jueces electrónicos?. In *Derecho penal y Comportamiento Humano. Avances desde la Neurociencia y la Inteligencia Artificial*. Tirant lo Blanch.

Patterson, D. (2019). Neuroscience and the explanation of human action. *Neurolaw and Responsibility for Action* (pp. 11–27). Cambridge University Press.

Pazouki, S., Golilarz, N.A., Kazemi-Razi, M., & Aydeger, A. (2022). *A self-healing cybersecurity mechanism for cyberattacks targeting artificial neural network-based human brain implants controlling smart homes*. In Proceedings of the International Conference on Electrical, *Computer and Energy Technologies ICECET* (pp. 1–6).

Peralta, J. M. (2012). *Motivos reprochables. Una investigación acerca de las motivaciones individuales para el Derecho penal liberal*. Marcial Pons, Madrid.

Pérez Manzano, M. (2013). El tiempo de la consciencia y la libertad de decisión: Bases para una reflexión sobre Neurociencia y responsabilidad penal. In *Neurociencias y Derecho penal. Nuevas perspectivas en el ámbito de la culpabilidad y tratamiento jurídico-penal de la peligrosidad* (pp. 105–135). BdeF / EDISOFER, Buenos Aires/Madrid.

Pérez Manzano, M. (2021). Algunos datos empíricos sobre la atribución de estados mentales. *Revista Electrónica De Ciencia Penal y Criminología, 23–15*, 1–22.

Pérez Manzano, M. (2022). Reflexiones sobre el concepto de dolo a la luz de los desarrollos de la neurociencia cognitiva. In *derecho penal y comportamiento humano. Avances desde la neurociencia y la inteligencia artificial* (pp. 393–416). Tirant lo Blanch.

Prinz, W. (2004). Was wissen und können Hirnforscher in zehn Jahren?. *Gehirn & Geist* , *6*, 31–37.

Pugh, J., et al. (2018). Brainjacking in deep brain stimulation and autonomy. *Ethics and Information Technology, 20*, 219–232.

Pycroft, L., et al. (2016). Brainjacking: Implant security issues in invasive neuromodulation. *World Neurosurgery*, 454–462.

Reche Tello, N. (2024). *Mens Iura Fundamentalia: La Neurotecnología ante la Constitución*. Colex, Madrid.

Rösler, F. (2004). Es gibt Grenzen der Erkenntnis—auch für die Hirnforschung. *Gehirn& Geist, 6*, 31–37.

Ryberg, J. (2020). Risk-based sentencing and predictive accuracy. *Ethical Theory and Moral Practice, 23*, 251–263.

Wajnerman, A., & López-Silva, P. (2022). Mental privacy and neuroprotection. A open debate. In *Protecting the mind. Challenges in law, neuroprotection, and neurorights* (pp. 141–156). Springer.

Yuste R (2019) *Las nuevas Neurotecnologías y su impacto en la Ciencia, Medicina y Sociedad* (pp. 41–53). Fundación Ramón Areces.

Yuste, R. (2023). Advocating for neurodata privacy and neurotechnology regulation. *Nature Protocols, 18*, 2869–2875.

Yuste, R., & de la Quadra-Salcedo, T. (2023). Neuro-rights and new charts of digital rights: A dialogue. Beyond the limits of the law. *Indiana Journal of Global Legal Studies, 30*, 15–37.

Yuste, R., Genser, J., & Herrmann, S. (2021). It's time for neuro-rights: New human rights for the age of neurotechnology. *Horizons: Journal of International Relations and Sustainable Development, 18*, 154–165.

Personal Identity and Personality Neurorights: Reconceptualize to Protect

José Angel Marinaro

Abstract The primary objective of this chapter is to establish links between the legal concepts of identity and personality. In the literature, both concepts are usually treated collectively, separately, or indistinctly, given that the neurotechnological mechanisms of affectation are identical. Since we are guided by a regulatory purpose by using the theoretical components of our dogma of reconceptualization, we are not indifferent to their contents. Establishing these links will be the basis of a schematization where we will observe that personality in its semantic field has the aptitude to incorporate the notion of (personal) identity. In turn, this will allow us to classify and systematically place psychological continuity and its close relative, identity self-perception.

Keywords Personal identity · Personality · Reconceptualization · Identity self-perception · Protection · Neurorights

1 Introduction

The process of reconceptualization of legal regulations requires the establishment of semantic consensus in regulatory function. This could help us to overcome the logical overlapping between different neurorights currently being proposed in the literature. Conceptual discrimination and their taxonomic schematization will facilitate the judicial defense required, for instance, by the "Habeas Cogitationem" as a means of expedited judicial protection recommended by UNESCO (2023, 10). In the absence of these considerations, the risk of rights inflation will arise. It will be imperative to determine the pre-existence of a legal text that indisputably (semantically and normatively) embodies the right to identity, surpassing the "filters" of eventual inconsistencies, contradictions or normative redundancies (Muñoz & Marinaro, 2022; Muñoz (2023) the same applies to the notion of personality and its eventual presence in the supranational texts under consideration.

J. A. Marinaro (✉)
Universidad Nacional de La Matanza, San Justo, Argentina
e-mail: marinarojoseangel@gmail.com

P. López-Silva (ed.), *Contextualizing Neuroprotection: Latin American Perspectives on the Impact of Neurotechnological Development in Life and Society*, Ethics of Science and Technology Assessment 54, https://doi.org/10.1007/978-3-031-96055-0_4

Following the dogmatic system of reconceptualization that we have conceived with José Manuel Muñoz (2023) and which has been referenced in the Council of Europe's Bertoni & Ienca report (2024), we shall review the expressions "identity" and "personality" from a semantic point of view, which prima facie, appears to be of interest in a regulatory function. In this context, we aspire to determine the role of the so-called "psychological continuity", apparently ascribed to the nomenclatures "identity" and "personality". Likewise, we will take a similar approach to "identity self-perception".

Only through the strict application of this dogmatic model of unquestionable philosophical, semantic and juridical character, shall we be in a position to answer the question: are the Rights to identity or personality a sound case of a normative gap that would lead us to assert the discovery of a genuine new neuroright? Moreover, could a category with distinctive nomenclature -and contents-appear on the scene? Is the identity self-perception, previously mentioned in a document of the Parlatino (2022), such a category?

This is a typical task of the "neurojurist" (Muñoz & Marinaro, 2022) as a circumstantial nexus agent between the interdisciplinary contributions of neurosciences "stricto sensu" and the areas of Law. A justified and temporary exception to Morse's claim of "neuromodesty", which rather seeks to set limits to the "invasion" of neurosciences, to which we certainly agree. Following the taxonomies in vogue (Ienca & Andorno, 2017; Yuste et al., 2017), we shall propose the reconceptualizing alternatives designed for the protection of psychological continuity from the category of personality.

2 Personality, Personal Identity, Psychological Continuity and Identity Self-perception

2.1 Voices of the Doctrine

The path we will follow attempts to establish that in the emerging doctrine there is an argumentative development that, relevant to the objective of this work, encompasses the notions of identity and personality which are addressed in an indistinct, unified and, sometimes, differentiated way without investigating their possible links from their corresponding scientific relevance. Note that the neurotechnological means by which the affectations of neurointerests are enabled are the same. The objective in this section is to highlight these claims in order to visualize the current regulations that outline both concepts.

From the notion of "psychological continuity", Ienca and Andorno (2017) note that:

> The right to psychological continuity can be seen as a special form of the right to identity resulting from neurotechnologies... the purpose of the right to psychological continuity is not to avoid the unrestricted access to brain information, but the induced alteration of neuronal

functionality...The Universal Declaration of Human Rights also addresses the right to have and develop a personality (art. 22).

According to Mănuc (2012), personality rights can be defined as those "that embody the quintessence of the human person, and are intrinsic to the human being" (Page).

Regarding the concept of private life, Susie Alegre (2017) thinks similarly when she argues that, "my right to a private life grants me a space in which I can develop my personality and practice being myself" (Page). From Australia and in legal and ethical-philosophical insights Nicole Vincent (2014) emphasizes, "Growing empirical evidence supports concerns that direct brain interventions may have adverse effects on such things as authenticity and personal identity by significantly altering character and personality" (Page). Critically, Bublitz (2022a, 2022b) stated:

> The right to personal identity is vague. What is the meaning of identity here? As bioethical debates have shown, the term is ambiguous... it may refer to diachronic identity (conditions under which a person is the same at two different periods), to personality (character), to group membership as per identity politics, and so on... it is unclear what might amount to a right to personal identity in the first sense... a right to one's personality in the second sense is clearer; it could protect against interference with one's personality, such as reputation or character. This is a valuable right, and it is, by this name or another, wholly or partially, recognized in existing law. Generally speaking, many continental European legal systems tend to dwell on personality rights, while U.S. law focuses on privacy rights... but from different perspectives, they both capture a wide range of interferences with personality (Page).

The scenario described by Leuenberger (2021) from his philosophical perspective is also based on the appreciation that neural interventions tend to provoke profound changes in personality, even with elements linked to arguments on identity, especially narrative, which, in other proposals, are attributed to alterations on (personal) identity. The Council of Europe itself poses in perspective possible forms of affectation of neurotechnologies on the notion of personality, fostering evaluations that, among other neurorights of the taxonomy in discussion, refer to psychological continuity. In its Strategic Action Plan on human rights and technologies in biomedicine (2020–2025), they stated that:

> Applications in the field of neurotechnology raise issues of privacy, personality and discrimination. Therefore, it is necessary to assess whether these issues can be sufficiently addressed within the existing human rights framework or whether new human rights related to cognitive freedom, mental privacy, mental integrity and psychological continuity need to be considered to guide neurotechnologies... (Page)

The work of de Haan et al. (2017), in philosophical perspective, explores whether treatments with DBS have the ability to provoke effects on the personality of patients. That is to say, the same neurotechnology that is mentioned by other authors in terms of alterations of identity or psychological continuity, according to de Haan et al. (2017) is oriented to eventual modifications on the personality considered also as modifications of the self.

In fact, an introductory subtitle groups these concepts by stating it in these terms: "Different definitions of changes in personality, in the self or in personal identity".

Referring to Schechtman (2010) de Haan considers the changes introduced or generated by DBS in Parkinson's disease as modifications of the personality to further address the reference that the cited author makes concerning the narrative vision on identity, when describing the threats to which it is exposed by the implementation of DBS. De Haan et al (2017) also raises questions analogous to those we outline here. She warns of the difficulties by explaining that: "The concept of 'personality' is not so easy to define, nor are the related concepts of 'self' and 'personal identity', and whilst he regards them as having different meanings, has decidedly treated them indistinctly" (Page). The undifferentiated approach to the identity and personality concepts also appears in several paragraphs of the renowned publication by Klaming and Haselager (2013).

In an interesting 2020 paper, the Oxford ethicist Jonathan Pugh addresses his reflections towards the need to appreciate the normative sense of the concept of personality, but lacks further explanation (even beyond its possible obviousness), to facilitate the comprehension of the norm and how it differs or resembles statements about identity. We neurojurists have a possibly divergent view of what we call "normative", closely related to "regulatory", perhaps a subspecies of the former. And that is why we find it unsettling to "put on the display" the neurotechnological advances (some of which refer to other species) that have a bearing on the contents of the PIAAAS, which refers to concepts such as personality, identity, autonomy, authenticity, agency and self and, on the basis of this frame of mind, to advance in the design of deontic precepts. These are what we call, alongside Jose Manuel Muñoz (2023), sections of reconceptualization of teleology or regulatory function. Bear in mind that the scenario of possible alterations is defined by Pugh in elements of the acronym PIAAAS, all of which are susceptible to diverse affectations due to the use of Deep Brain Stimulation as shown by empirical surveys.[1]

The jurist Sjors Ligthart (2024) recently addressed the rarely discussed issue of "psychological continuity". The author claimed that there is no need for the creation of a new neuroright concerning psychological continuity because, according to his assertions, such a right is already encompassed by the rights to personal identity, self-determination and personal integrity. In a way, Ligthart is reconceptualizing these statements, yet he does not explain how he conducts this procedure. Moreover, such an interpretation does not exceed the level of relocations in the doctrinal field without resorting to the search for accommodation in current regulatory norms. In the end, reconceptualizing is a coherent shortcut that avoids the inflationary risks of the variant on the creation of new neurorights. According to the quotations from the "neurodoctrine"[2] that we have briefly described in this passage, it is clear that the consideration of the right to "personal identity" and the consequent right to "personality" has been of interest all along, without any attempt to establish a semantic axiological nexus between the two that would open the way to possible regulations in a reconceptualizing scheme.

---

[1] Do not exclude Glannon (2009); Müller et al. (2015); Leentjens et al, (2004).

[2] Neologism introduced by the author which refers to the subspecies of the legal doctrine that restricts the field of analysis to the neurobiological implications affecting neurorights.

However, by examining some provisions of current international law, we will be able to take steps to establish the interconnection that we propose as a variant to other proposals.

3 Exploring the Legal Texts that Mention the Right to Identity: In Search of the Normative Space

Our goal, we insist, is primarily to reconceptualize, an idea primarily introduced by Ienca and Andorno (2017) and then structured, developed and expanded by Muñoz and Marinaro as recognized by Bertoni and Ienca (2024); and thus the dogmatic task to be undertaken (Muñoz et al. 2023, Muñoz & Marinaro 2022) requires the revision of the existing regulations. A search for nominative identities that will later allow a semiotic exegesis of their material contents in order to locate the contents of the semantic sense fields that must be re-signified.

In this case, we will look at the external side of the States (supranational texts) in search of those provisions that have the capacity to cover the proposed taxonomies. We will begin with the Convention on the Child's Rights, which has regulations of great value for the purpose of this work. In some provisions, we find the desired statement: articles 8, 11, 12 and 13 on the right to identity in the registry variant.

We find in art. 29 sub "c" the explicit allusion to cultural identity, a normative reference that is closely related to our idea of identity self-perception, which, in addition, has distinctive features in respect to psychological continuity for the reasons that we have extensively provided on another editorial occasion (Muñoz & Marinaro, 2022). As noted by Ienca and Andorno (2017), the European Court of Human Rights (ECHR or The Court)[3] has recognized the right to identity under the general right of respect for private life in accordance with Article 8 ECHR. The aforementioned interpretation can be considered as a reconceptualization of the right to identity contained within or "under the general right" of respect for private life.

However, let us pay attention: according to the ECHR, the notion of private life encompasses the "right to identity" and the "right to personal development", whether in terms of personality or personal autonomy.

We believe that even if other interpretations derive personal identity from private life, these possible overlaps-as we mentioned-should not confuse us. We will contend that identity in its genesis has given way to the notion of personality and it is there, for reasons of evolutionary connection, where its normative space resides.

Ligthart (2024) himself regarding the European framework and psychological continuity alludes to some general notions and concepts, such as "mental stability", "identity" and "personality", pointing out that they are still poorly defined in the case law of the ECHR. In this regard, these statements also establish links between identity and personality, even though they do not specify or develop the contents of these links.

[3] Goodwin v. the United Kingdom ECtHR (2002) 35, 18–90.

In the following paragraph by the Dutch professor, we can see some resemblance to the thesis we postulate here: "Cases such as these illustrate that the introduction of new technology is sometimes accompanied by new challenges for the shaping and development of personality and identity" (Ligthart, 2024, Page). Thus, we see that the idea of "personality" is permanently present in various references to the notion of "identity".

These interpretations, surely for lack of anticipation in the European Convention, do not suffice to establish the links between identity and personality, which are only sporadically mentioned. In fact, this absence of reference to personality leads some scholars to "search" in the right to integrity for an alternative space of legal protection for psychological continuity (Ligthart, 2024) from where we find another assumption of reconceptualization—forced this time—put into praxis.

Possibly the confusion of overlaps in neurorights may lead to said alternatives.

The variant we propose, which extends its gaze to the UDHR, on the capacity to cover the right to the development of personality, taking into account the evolutionary cycle of the concept of identity that we will elaborate later on, is more akin to the objective of protecting the aforementioned right against the misuse of new technologies and their harmful capacity to intervene in the human brain and mind. We have established that material damage to the brain (for example, to neuronal tissue) already finds a protective framework in the current forms of the right to the preservation of physical integrity, including in the criminal rules of the internal side of the States that protect bodily integrity, which is the reason why this conduit is not suitable for the preservation of personal identity insofar as it is understood as a threat to mental states.

These "forms" by themselves pave the way to understanding the nature of the affectations of identity/personality; as in these spaces, distinctions such as psychological continuity and identity self-perception come into play. In either case, both the psychological and cultural effects are a matter of specific protection.

Whereas damage to mental integrity refers, in terms of its purpose of protection, to the modification of neuronal activity, this may or may not subsequently lead to the affectation of personality traits and may only be limited to the modification of neuronal computation and/or to means of manipulation that do not alter identity. In order to emphasize the difference, the development of varied memory engineering techniques are inherent challenges to mental integrity (Ienca & Andorno, 2017).

The singularities of "mental damage" allow us to differentiate the specific damages of integrity from those that might have those signs resulting in affectations that are not of punctual neuronal computations but of mental states whose greater presence are the continuities of certain personality characteristics that enable the sustainability of psychological and/or anthropological-cultural traits through which people "identify" themselves.

Those who have suffered ostracism and exile can provide an account of these forms of affectation of traits inherent to what we refer to as identity self-perception, which is vulnerable to disruption with the use of some neurotechnologies. These examples, which are beyond neuroaffectations, are also indicative of interruptions of the "narrative flow" (Baylis, 2013) derived from traumatic events or experiences with possible influence, in some cases, on cultural identity. In his criticism of the jurisprudence of the Court on psychological continuity Ligthart (2024) states that:

> It refers to the right to psychological, psychic and moral integrity without clarifying the meaning, scope and distinctive functions of these seemingly different rights. Likewise, it develops and applies, case by case, a right to identity and personality, addressing the importance of preserving mental stability, albeit without defining these notions and clarifying a fair understanding of requirements for identity and personality (Page 31).

The Court has also explored the semantic space of personality without providing delimitations that would make it possible to discern the spheres of protection involved. In the search for the right to personal identity, given the apparent link with the idea of personality mentioned by Ienca and Andorno (2017), we find that some international instruments mention two variables: personality and legal personality. It seems clear that these are distinct situations.

4 Brief Exploration of the Legal Texts that Mention the Right to Legal Personality.

Regarding the regulatory mentions of the idea of "legal personality", and without pretending to exhaust the investigations, we see that the International Covenant on Civil and Political Rights in its art. 16 expressly states: "Everyone shall have the right to recognition everywhere as a person before the law".

Likewise, in almost identical wording, Article 3 of the American Convention on Human Rights, preceded by an introductory paragraph stipulating the right to recognition as a person before the law, provides that: "Every person has the right to recognition as a person before the law". The mention of the concept of "personality" in the Universal Declaration of Human Rights is included in several passages under the idea of the right to the development of personality. In its article 22, we read that:

Everyone, as a member of society, has the right to social security and is entitled to realization, through national effort and international co-operation and in accordance with the organization and resources of each State, of the economic, social and cultural rights indispensable for his dignity and the free development of his personality.

Subsequently, with regard to the right to education, Article 26, sub 2, is written as follows:

Education shall be directed to the full development of the human personality and to the strengthening of respect for human rights and fundamental freedoms; it shall promote understanding, tolerance and friendship among all nations, racial or religious groups, and shall foster the development of the activities of the United Nations for the preservation of peace.

Finally, from a deontic point of view, in its article 29, sub 1: "Everyone has duties to the community in which alone the free and full development of his personality is possible".

5 Identity and Personality. The Need to Determine Their Links for the Task of Reconceptualization

In this section, having already highlighted the recurrent undifferentiated quotation of these concepts, let us move on to the development of the reasons for which we consider that it would be plausible to establish the links between both, thus enabling the work of reconceptualization in a regulatory function. In the presence of this picture, a consensual variant emerges, and we can say that the personality (plainly) that appears in the quoted texts, is a generic concept that embodies the conceptualizations and interpretations on the notion of identity and reconceptualize on that basis in order to evolve said statement.

It is necessary to establish the links between personality and identity, since with the exception of the UDHR and the cultural or anthropological scope; there are no "soft law" regulations that embody the right that we perceive as personal identity. Since there are mentions of the right to personality (in general, its development) separate from its "legal" species, we must also clarify the references to "registry" type identity, as they do not concern affectable states of mind.

If we reach the conclusion that personality is an all-encompassing concept of identity, the task of redefining that which is contained in supranational norms emerges as a possible axiology. We outline a different or alternative path to what Ligthart (2024) proposed, as he mainly examines some European legal texts without taking into account the provisions of the UDHR, to later infer "how the right to psychological continuity is related to and covered by, respectively, (1) the right to personal identity, (2) the right to self-determination, and (3) the right to personal integrity" (Page 6).

Let us make the attempt to support our choice by clearing some terms: Manuc (2012) provides a broad notion by aptly differentiating legal personality from personality, where the former is a reduced form of the broader latter: "Personality... is not limited only to the technical concept of legal personality, rather it encompasses the whole human person, in its biological, psychological and social reality" (Page 360/370).

When alluding to the legal capacity of personality, Bublitz (2022a, 2022b) highlights a set of:

Factual mental properties such as the ability to reason, to appreciate moral and legal norms, to understand the situation one faces, one's options and their possible consequences, a structure of preferences; powers of self-control and a basic level of mental stability; memory and other abilities to form, reflect and express one's will, all of these in the absence of factors that undermine them, such as serious mental disorders. In essence, being a person requires a set of mental properties (Page 3).

For such reasons we believe that all of them are susceptible to be affected and at the same time are indicative from this perspective, of the set of additional aspects contained in such notions such as identity in the broad and diversified sense that we propose here for the task of reconceptualizing.

6 We Delve into the Genesis of Psychology and Social Psychology. An Insightful Path from Philosophical and Psychological Basis

In the roots of both individual and social psychology of the notion of personal identity, we may have to backtrack to the forms of the self, our "own self" (Freud, 1921) and its development to realize that these origins reveal an unbreakable link with the concept of "personality". The phenomenological perspective (Gallagher & Zahavi, 2014), also recommended by Bublitz (2023), offers solid and insightful analyses of self-experience and the different concepts of the self that can be informative for the recent interest in these issues in cognitive science. Specifically, phenomenology shows that the "self" is significantly implicated in all aspects of experience, including phenomenality, temporality, corporeality, action, and our interaction with others. Does this refer to narrative identity?

Searle's philosophy (1996) seems to reach similar conclusions from the elaboration of concepts such as subjectivity and consciousness. The conscious mental processes and states find their distinctive trait in subjectivity. The possibilities of self-perception lay the ground for notions of identity and personality. But the consolidation of convincing links between the concepts of identity and personality, the lack of which we have been claiming in the literature, appears solidly elaborated by Elisa Fernandez (2012), in her study for the Spanish Society of Psychosomatic Medicine and Psychotherapy, which offers the best explanation by affirming that the pairs identity and personality are two concepts that go together: one is and differs from others by this awareness of being unique and different from the rest.

Hence, we conceptualize personal identity as the set of both internal and external properties that define someone as a non-multiple individual entity that makes someone who they are, and distinguishes them from others. Delving into the perspective of historical development Fernandez (2012) adds that:

The concept of 'identity' was originally studied from a psychoanalytical perspective, from the 'I' as an integrating force of the human being. The human being, as an individual facing the pressure of social norms. Subsequently, thanks to Jung, among other authors, the concept of "identity" was linked to that of "the mass", or society, being observed and studied from the emerging "Social Psychology". In the 20th century another concept was again studied, personality, from the individual as a pattern of psychological characteristics that are manifested in behaviors sustained over time, which differentiate him/her from others... (Page 1).

A hint of a connection between the concepts of identity and personality appears in Ienca (2021):

Personal identity is known as the set of properties that define someone as an individual person or make someone the person they are and that separates them from others... consequently the notion of personal identity often implies a notion of personality, i.e. the state of being a person as opposed to a non-person... personality as a set of mental properties... (Page 5).

Although he does not delve into the further foundations of this linkage, one can say that Ienca (similarly to Bublitz) leaves the way open for establishing its concrete contents, in the way we do here.

For those of us who ponder from semiotics and the spheres for the meaning of Law, we find it extremely useful when considering derived conceptualizations, to keep in mind that the transition towards subsequent explanations of preferential narrative forms of identity (Baylis, 2013) with psychological links or connections (Parfit, 1984) as a necessary but insufficient condition, requires some drafts of the planes of consciousness and the relationships of my subjective and inter-subjective "I" where we can incorporate the enactivist cognition of the Berkeley philosopher Alva Noë (2006) which integrates the perception of the world and so our mind is not a mere data processor. All of its implications cannot be dealt with here.

In this interdisciplinary discourse we resort to social psychology, neurobiology, and philosophy of mind, for the task of semantic consensuses in regulatory function that, as mentioned, focuses on the deontic in order to determine actions or abstentions. The issue is, on the one hand, not to confine the concepts exclusively to the intricate (and growing) field of theories of identity linked to "controversial psychological theories" originating from neuroethical debates (Pugh, 2020) and, on the other hand, the need to be clear given the normative implications. Neurojurists must take these precautions. Our holistic view clearly adds the normative teleological component.

Regulation is a complex task whose transcendent field is the interpretative task of the norms, both of "lege ferenda"-the law we claim- and of "lege lata"-the law as such-.

In view of what we have developed, all that remains is to ask ourselves what kind of conceptual relationship we can glimpse between identity and personality in order to advance in a valuative management that ensures the coherence of its result.

7 Framing the Proposal

Hereby, we propose that, with the aforementioned precautions, from the psychological and psychological-social genesis of both terms we can establish (reach a consensus) that the right to the development of the personality has a "genus proximus" relationship with the notion of legal personality and personal identity in whose content we find both the psychological continuity and the identitarian self-perception, as values of neuroprotection. The so-called registry identity clearly refers to other aspects beyond the tasks herein developed.

In this case, both (psychological continuity and identity self-perception) aim to avoid not the unrestricted access to brain information (which is inherent to Mental Privacy (Ienca & Andorno, 2017) but the induced alteration of neuronal functioning as such, which involves, by means of these maneuvers, the alterations of both fields of personal identity. In this scheme the infringements on the cultural identity of N. N. and A. of article 29, at. "c" of the CRC, shall be covered.

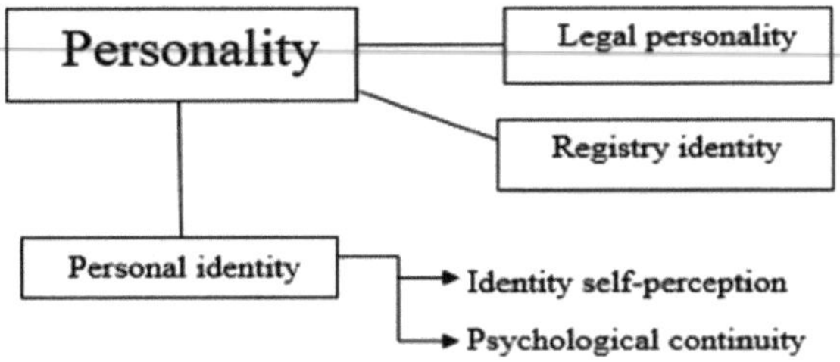

Fig. 1 Author's proposal of the map of the content of 'personality'

The 2023 UNESCO Report for Latin America and the Caribbean also emphasizes in its Principle No. 5 the injurious ways and means by stating that the goal is to ensure that personal identity is not deliberately altered by third parties through neurodevices without the free and well-informed consent of the person concerned.

The 2023 Declaration of Principles of the IJC (OAS) provides a notion of personal identity which adds similar contents to those we propose. By conceptually developing Principle 1, on identity, autonomy and privacy of neuronal activity, it establishes that "personal identity is a human right that is understood as a highly complex concept, inextricably linked to the self-perception of the personality and comprising anthropological, cultural and social elements that underlie the individuality and true identity of a person".[4]

In this normative framework, we can say that there is an advance towards a possible and coherent reconceptualization of the right to personality embodied in the aforementioned supranational norms. Hence, we shall arrive at the following classification (Fig. 1):

8 Final Considerations

Providing that there is always a sense of the regulatory purposes already mentioned, which seeks semantic consensus, our proposals for the definition of the concepts in contention lead us to understand that personal identity refers to a set of both internal and external properties that define someone as a non-multiple individual entity that makes such a person be what he or she is, and is what distinguishes him or her from others.

On the other hand, in addition to the psychological continuity already discussed and elaborated by Klaming and Hasselager (2013) and Ienca and Andorno (2017), we understand that identity self-perception is a concept linked to psychological continuity, referring existentially to the identity of a person in which a variety of factors beyond the strictly psychological ones (namely social, ideological, religious and

[4] As a distinctive detail, we note that the General Bill on Neurorights and nuerotechnologies submitted to the Senate of Mexico (2024), introduces a criminal offense (art. 347) that penalizes the illicit manipulation of Personality.

cultural) intervene. This is an eminently dynamic constitutive element, of progressive self-narrative conformation and subject to important changes in the environment that allow shaping the traits that individualize the cultural diversities of persons liable to neurotechnological disorders.

Finally, Fernandez (2012), like us, go one the viewpoint of Millon (1998), conceiving personality us a "complex pattern of deeply rooted psychological characteristics, mostly unconscious and difficult to change, that are expressed automatically in almost all areas of an individual's functioning" where identity is an inherent part of personality. (Page 7).

References

Alegre, S. (2017). Rethinking freedom of thought for the 21th Century, E.H.R.L.R., Issue 3 Thomson Reuters (Professional) UK Limited and Contributors.

Baylis, F. (2013). (2013) "I am who I am": On perceived threats to personal identity from deep brain stimulation. *Neuroethics, 6*(3), 513–526. https://doi.org/10.1007/s12152-011-9137-1

Bublitz, J. C. (2022). Might artificial intelligence become part of the person, and what are the key ethical and legal implications? *AI & SOCIETY, 39*, 1095–1106. https://doi.org/10.1007/s00146-022-01584-y

Bublitz, J. C. (2023). What an international declaration on neurotechnologies and human rights could look like: ideas. *Suggestions, Desiderata, A JOB Neuroscience, 15*, 96–112. https://doi.org/10.1080/21507740.2023.2270512)

Bublitz, J. C. (2022). Novel neurorights: From nonsense to substance. *Neuroethics, 15*(7)

Council of Europe. (2024). The privacy and data protection implication of the use of neurotechnology and neural data from the perspective of Convention 108, by Eduardo Bertoni and Marcello Ienca. https://globalfreedomofexpression.columbia.edu/wp-content/uploads/2024/08/Expert-Report-Neuroscience.docx.pdf

De Haan, S., Rietveld, E., Stokhof, M., & Denys, D. (2017). Becoming more yourself? Personality changes after DBS treatment for psychiatric disorders: Experiences of OCD patients and general considerations. *PLoS ONE, 12*(4), Article e0175748. https://doi.org/10.1371/journal.pone.0175748

Fernández, E. (2012). Identidad y Personalidad: o como sabemos que somos diferentes a los demás. y Sociedad Española de Medicina Psicosomática y Psicoterapia, vol. 2. Nro.4 https://www.psicociencias.org/pdf_noticias/Identidad_y_personalidad.pdf

Freud, S. (1921). Psicología de las masas y análisis del yo. Ed. Biblioteca Nueva. Tomo III. Obras Completas

Gallagher, S., & Zahavi, D. (2014). *The Phenomenological Mind*. Alianza Editorial. Madrid. Tr. Marta Jorba Grau. ISBN: 978-84-206-8931-9. Original Title: The Phenomenological Mind: An Introduction to Philosophy of Mind and Cognitive Science. Routledge, Taylor & Francis Group.

Glannon, W. (2009). Stimulating brains, altering minds. *Journal of Medical Ethics, 35*(5), 289–292. https://doi.org/10.1136/jme.2008.027789. PMID: 19407032.

Ienca, M. (2021). On Neurorights. *Frontiers in Human Neuroscience, 15*, Article 701258. https://doi.org/10.3389/fnhum.2021.701258

Ienca, M., & Andorno, R. (2017). Towards new human rights in the age of neuroscience and neurotechnology. *Life Sciences, Society and Policy, 13*(5), 1–27. https://doi.org/10.1186/s40504-017-0050-1

Klaming, L., & Haselager, P. (2013). Did my brain implant make me do it? Questions raised by DBS regarding psychological continuity, responsibility for action and mental competence. *Neuroethics, 6*(3), 527–539.

Draft General Law on NeuroRights and Neurotechnologies submitted to the Senate of Mexico. (July 17, 2024). https://www.senado.gob.mx/65/gaceta_comisión_permanente/documento/143375

Leentjens, A. F., Visser-Vandewalle, V., Temel, Y., & Verhey, F. R. (2004). Manipuleerbare wilsbekwaamheid: Een ethisch probleem bij elektrostimulatie van de nucleus subthalamicus voor ernstige ziekte van Parkinson [Manipulation of mental competence: An ethical problem in case of electrical stimulation of the subthalamic nucleus for severe Parkinson's disease]. *Nederlands Tijdschrift Voor Geneeskunde, 148*(28), 1394–1398.

Leuenberger, M. (2021). Losing meaning: Philosophical reflections on neural interventions and their influence on narrative identity. *Neuroethics, 14*, 491–505. https://doi.org/10.1007/s12152-021-09469-5-

Ligthart, S. (2024). Towards a human right to psychological continuity? Reflections on the rights to personal identity, self-determination and personal integrity?. In *The European Convention on Human Rights Law Review* https://doi.org/10.1163/26663236-bja10092

Mănuc, L. M. (2012). Features and evolution references to personalityrights. *Contemporary Readings in Law and Social Justice, 4*(1), 360–370.

Millon, T. (1998). Trastornos de la personalidad. Más allá del DSM-IV. Ed.Masson.

Müller, S. et al. (2015). Rivaling paradigms in psychiatric neurosurgery: Adjustability versus quick fix versus minimal-invasiveness. *Frontiers in Integrative Neuroscience, 9*, 27.

Muñoz, J. M. (2023). Neurorights as reconceptualized human rights. *Frontiers in Political Science, 5*, 1322922. https://doi.org/10.3389/fpos.2023.1322922

Muñoz, J. M., Marinaro, J.A. (2022). Neurorights: From novelty to reconceptualization. Rule of law agenda https://agendaestadodederecho.com/neuroderechos-de-la-novedad-a-la-reconceptualizacion/

Muñoz, J. M., & Marinaro, J. Á. (2024). "You shall have the thought": habeas cogitationem as a New Legal Remedy to Enforce Freedom of Thinking and Neurorights. *Neuroethics, 17*(1), 18. https://doi.org/10.1007/s12152-024-09551-8

Noë, A. (2006). *Action in perception.* The MIT Press.

OEA. Comité Jurídico Interamericano (2023). Declaración de Principios Interamericanos en materia de Neurociencias, Neurotecnologías y Derechos Humanos. Río de Janeiro, March 9, 2023. CJI/RES. 281 (CII-O/23) corr. 1.

Parfit, D. (1984). *Reasons and persons.* Clarendon Press.

Parlatino (Parlamento Latinoamericano y Caribeño). (2022). Declaration with recommendations on the need to introduce neuro-rights in the legislations of the Congresses of PARLATINO. https://parlatino.org/wp-content/uploads/2017/09/declaracion-neuroderechos.pdf

Pretty v. United Kingdom 2346/02 (ECHR, April 29, 2002). https://hudoc.echr.coe.int/eng?i=001-162224

Pught, J. (2020). Clarifying the normative meaning of "personality" Changes following deep brain stimulation. *Science and Engineering Ethics, 26*(1655), 1680. https://doi.org/10.1007/s11948-020-00207-3

Schechtman, M. (2010). Philosophical reflections on narrative and deep brain stimulation. *The Journal of Clinical Ethics., 21*(2), 133–9. https://doi.org/10.1086/JCE201021206

Searle, J. R. (1996). The rediscovery of the mind. The MIT Press, Cambridge, Massachusetts. ISBN 84-7423-742-4

UNESCO. (2023). UNESCO Office Montevideo and Regional Bureau for Science in Latin America and the Caribbean: Desafíos y propuestas de Política Pública, Oficina regional Montevideo. https://unesdoc.unesco.org/ark:/48223/pf000038

Vincent, N. (2014). Restoring responsibility: Promoting justice, therapy and reform through direct brain interventions. *Criminal Law and Philosophy, 8*, 21–42.

Yuste, R., Gatenby, J. C., Gerfen, C. R., Goering, S., Hawes, D., Holmes, P., & Yuste, R. (2017). Four ethical priorities for neurotechnologies and AI. *Nature, 551*(7679), 159–163. https://doi.org/10.1038/551159a

Deconstructing the Mind: Neurotechnology, Mental Privacy, and the Case for Neurorights

Raúl Madrid

Abstract This chapter explores the challenges posed by neurotechnology and artificial intelligence to traditional understandings of mental privacy and human rights, employing Jacques Derrida's deconstructive approach. Advances in neurotechnology, such as brain-computer interfaces (BCIs), functional magnetic resonance imaging (fMRI), and transcranial magnetic stimulation (TMS), have blurred the boundary between the "inside" (thoughts and mental states) and the "outside" (observable actions). These technologies enable unprecedented access to and manipulation of thoughts, emotions, and decisions, raising ethical concerns about privacy, autonomy, and free will. The chapter situates these issues within Derrida's critique of metaphysical oppositions, specifically the inside/outside dichotomy, and his concept of the "trace," which destabilizes binary distinctions. By analyzing the implications of thought as a readable "writing" within the brain, the chapter suggests that neurotechnology deconstructs traditional notions of the mind as an inviolable interior space. It argues that this transformation demands urgent legal responses, such as the incorporation of "neurorights" into constitutional and statutory frameworks to protect mental integrity. The discussion underscores the importance of a robust legal infrastructure to address risks including neuroprivacy violations, manipulation of free will, and neurodiscrimination, as well as the broader philosophical implications of rethinking the human mind as a material system open to external interpretation and influence.

Keywords Neurotechnology · Neurorights · Deconstruction · Mental privacy · Artificial intelligence

R. Madrid (✉)
Faculty of Law and Faculty of Philosophy, Law, Science and Technology Program, Pontificia Universidad Católica de Chile, Santiago, Chile
e-mail: rmadrid@uc.cl

P. López-Silva (ed.), *Contextualizing Neuroprotection: Latin American Perspectives on the Impact of Neurotechnological Development in Life and Society*, Ethics of Science and Technology Assessment 54, https://doi.org/10.1007/978-3-031-96055-0_5

1 Background

The debate over the need for mental protection has gained significant traction in both public and academic spheres, driven by the rapid development of technological tools enabling unprecedented access to the inner workings of human consciousness and thought. This heightened focus on safeguarding mental interests is further reinforced by the growing emphasis on human rights, as reflected in constitutional guarantees that extend to a broad spectrum of actions and omissions, even those not explicitly articulated within constitutional texts. One avenue opened by technological advancements in clarifying such guarantees is the emergence of so-called "Neurorights". When the results of the Brain Activity Map Project (BAMP), a component of The Brain Initiative, were published—mapping the pathways that electrical impulses follow to produce states such as thoughts, perceptions, desires, beliefs, and voluntary or involuntary bodily movements—the scientific community realized that therapeutic applications would not be the sole use of this technology (Yuste et al., 2017). This realization underscored the necessity for ethical and regulatory scrutiny.

In response, a three-day workshop at Columbia University convened leading academics to explore the ethical implications of neurotechnology and artificial intelligence. The workshop gave rise to the concept of "Neurorights" and the NeuroRights Initiative (NRI) at Columbia University. Since then, debates over the definition, scope, foundation, and legal status of this novel framework of human rights have intensified. This chapter seeks to provide a fresh perspective on neurorights by leveraging Jacques Derrida's philosophy to examine the blurring of the boundary between what is "inside" and "outside" the mind. This shift underscores the pressing need to protect the identity, privacy, and freedom of our mental states.

2 The Eclipse of the Exteriority/interiority Opposition

A central premise in Derrida's philosophy is that every logocentric or metaphysical epoch conceptualizes reality as structured through pairs, each term defined in relation to the other, with both terms existing in mutual externality. These pairs are fundamentally and radically opposed, excluding one another by their very essence. Such terms are considered "pure," as exemplified by the primary metaphysical pair: *being* and *nothingness*, where nothing of *being* is present in *nothingness*, and vice versa. This radical opposition underpins the principle of non-contradiction, as articulated by Aristotle in *Metaphysics* (2016, II. 2). Derrida argues that this mode of binary thinking is not limited to classical or medieval philosophy but also characterizes modern thought, which, while formally adhering to the same structure, often inverts the content. In Derrida's view, this "law of mutual exteriority" applies universally to all such oppositional pairs, making them inherently distinguishable by their defining principles.

Derrida's analysis, however, does not primarily focus on the opposition between *being* and *nothingness*. Instead, he examines another foundational dichotomy in Western philosophy: *speech* and *writing*. In *Origin of Geometry: An Introduction* (Derrida, 1989), Derrida highlights the essential role of writing in transmitting geometric knowledge, enabling discoveries to persist and circulate beyond their original context. However, he critiques Edmund Husserl's phenomenology for its tension between original intuition and written expression. Derrida observes that Husserl, in *Logical Investigations*, fails to fully account for how writing impacts and transforms the purity of geometric intuition.

In *The Voice and the Phenomenon* (2011), Derrida challenges Husserl's privileging of presence, wherein meaning is tied to the immediate presence of an object and language is seen as merely representing that presence. Derrida argues that meaning does not rely on this immediacy. For Husserl, spoken language enables self-presence—when speaking, one hears one's own words, confirming the presence of both thoughts and self. Derrida disputes this notion, asserting that even spoken language is marked by *différance*—a process of both difference and deferral—that disrupts full presence.

Contrary to the traditional privileging of speech over writing—associating speech with immediacy and writing with absence—Derrida contends that writing is fundamental to the very possibility of language. Writing is not secondary or derivative; instead, speech already contains the characteristics of writing, as both are shaped by *différance*.

Derrida consolidates these ideas in his seminal work *Of Grammatology* (1967), where he further explores the interior/exterior dichotomy through the concept of a universal *grama* (writing). He parallels the dichotomies of speech/writing and inside/outside, questioning whether writing can truly be excluded from the interiority of language. In what Derrida terms the "representationalist position" (i.e., metaphysics), writing is treated merely as a representation of language, external to its "internal system" and secondary to speech. Ferdinand de Saussure reinforces this view in *Course in General Linguistics* (Saussure 2013), where he argues that writing is unrelated to the intrinsic system of language. For Saussure, the linguistic sign is arbitrary, with no inherent connection between the signifier (the form, such as sounds or letters) and the signified (the concept or idea). Consequently, writing is seen as a tool for preserving meaning, not as an integral part of language itself.

However, Derrida points out that Saussure himself acknowledges writing's inescapable influence on language, noting that it cannot be entirely dismissed. Derrida asks:

> Why else would [Saussure] give so much attention to that external phenomenon, that exiled figuration, that outside, that double? Why does he judge it impossible 'to simply disregard' what is nevertheless designated as the abstract itself with respect to the inside of language? (Derrida 1997, 34)

To address this ambiguity, Derrida calls for a "general linguistics" encompassing both speech and writing. He introduces *grammatology*, a deconstructive critique of metaphysical assumptions in thinkers like Plato, Saussure, and Rousseau. Writing, for Derrida, is not merely a representation of speech but a transcendental condition for the possibility of language. It serves as a *trace*, an entity that disrupts the boundaries between signifier and signified and exposes the interdependence of all linguistic elements.

This perspective undermines the notion of a clear distinction between interior and exterior. The traditional view of this dichotomy as natural and universal becomes, for Derrida, a construct subject to critique. By revealing how writing is intertwined with speech, he challenges the logocentric assumption that language is complete within itself, free from external influence. This deconstruction extends beyond linguistics, serving as a broader critique of binary oppositions in Western thought.

Derrida observes that even Saussure, who attempted to marginalize writing as external to the inner system of language, implicitly acknowledges its unavoidable role:

> Why else would [Saussure] give so much attention to that external phenomenon, that exiled figuration, that outside, that double? Why does he judge it impossible 'to simply disregard' [literally 'make abstraction of'] what is nevertheless designated as the abstract itself with respect to the inside of language? (*Of Grammatology* 1967/1997)

To resolve this constitutive ambiguity, he proposes the necessity of a general linguistics—what he terms "grammatology" or the general science of writing. Grammatology serves as a deconstructive critique of the metaphysical assumptions inherent in the works of Plato, Saussure, and Rousseau.

This way, writing turns out to be not merely a derivative representation of speech but constitutes a transcendental layer—a *trace*—that underpins the possibility of the empirical-transcendental distinction itself. Writing, in this sense, replaces the metaphysical entity (*ens*) as the primary object of thought. As Derrida famously asserts, "the trace must be thought before the entity" (*Of Grammatology* 1967/1997). This idea forms the basis of his deconstructive method: if all signs are unmotivated institutions, then no hierarchy or subordination exists among signifiers or among orders of signification. Writing, understood as *inscription*, envelops the entire linguistic field, dissolving distinctions between primary and secondary forms of meaning.

This thesis inherently challenges the conventional distinction between the interior and exterior. At first glance, the difference between what is "inside" and "outside" may seem intuitive, especially in the tangible world. Traditionally, this distinction has been regarded as transcendental—natural, universal, and intrinsic to the definition of paired opposites. Such a dichotomy has been central to metaphysical thought. Over time, however, Derrida argues that this separation transforms into a complex interplay of shifting relationships, ultimately displacing its claim to any foundational basis.

Derrida's argument unfolds as follows: the supposed transcendental difference between inside and outside is not a static ontological truth but a demand of thought itself, an assumption particularly evident in art. Here, discourse shifts from limits to frames; for instance, the frame of a painting acts as a boundary, structuring how the artwork is perceived and discussed (*The Truth in Painting* 1987). This act of distinguishing between interior and exterior is readily apparent in visual art but operates more subtly in other domains. Derrida contends that the function of the frame (or limit) is aligned with a persistent, transcendental need to delineate the two realms—a need he sets out to dismantle.

This critique brings Derrida to the concept of the *parergon*. The integrity and originality of an artwork depend on the supplementary nature of the *parergon*, which refers to elements that, while not essential to the artwork, are nonetheless integral to its interpretation and appreciation. These include the frame, lighting, placement, and other surrounding features. While traditionally viewed as external or ornamental, such elements influence how the artwork is perceived, challenging its supposed autonomy. The term originates from Kant's distinction between *ergon* (the work) and *parergon* (the hors d'oeuvre or ornament), as outlined in *Critique of Judgment* (Kant, 2000). Derrida expands on this concept, demonstrating that the *parergon* destabilizes the boundary between what is intrinsic to the artwork and what lies outside it.

In Derrida's framework, the *parergon* assumes profound significance as it blurs the dividing line between the artwork and its environment, questioning the strict criteria that separate inside from outside (*The Truth in Painting* 1987, 29). The figure of the *parergon* extends beyond art to encompass texts and symbolic systems, applying to any set of oppositions a text might contain. This notion imbues the interior/exterior dichotomy with a transcendental quality, making it a cornerstone of deconstruction. A *parergonic* or deconstructive reading refuses to acknowledge a fixed central meaning. Instead, it erases boundaries between interior and exterior, dismantling the self-identity of the inside by incorporating what was previously considered an external addition into the object's totality (Derrida, 1987, 64).

This strategy resists systematic structures, geometric order, or any transcendental meaning that can be separated from the interplay of signifiers. Instead, it "forces the system within the system itself" (Peñalver, 1995, 16). The crux of Derrida's argument lies in the *indecidability* of the boundary: the *parergon* occupies a liminal space, neither entirely inside nor outside. This unresolved ambiguity exemplifies deconstruction's broader aim: to expose the infinite texture, trace, and precariousness of a text, reformulating meaning in a dynamic, *poietic* sense.

In a broader sense, the *parergon* encapsulates the impossibility of a strict exteriority between opposites. It denies the notion of contradiction between the supplementary and the artwork, and more fundamentally, between inside and outside. This negation takes on a transcendental dimension, challenging the metaphysical and logocentric frameworks that rely on such oppositions. By interrogating these boundaries, Derrida redefines the conditions of meaning and representation, demonstrating that no element—be it within or beyond a system—exists in isolation. Instead, all elements are interwoven in a network of relations that continually destabilize their purported autonomy.

3 Neurotechnology as a Boundary Closer

The inside/outside dichotomy was not merely a fundamental construct of metaphysical and modern thought; it also provided a cognitive framework applicable to a range of practical and theoretical domains. This dichotomy facilitated a reliable conceptual operation by distinguishing between two opposing realms. One of the most apparent applications of this separation was the differentiation between internal processes, such as thought and decision-making, and external manifestations, such as physical movements and actions in the corporeal world.

This distinction is now being fundamentally challenged by advancements in neurotechnology and artificial intelligence, which suggest the possibility that third parties might gain access to an individual's internal thoughts and exert influence over their mental and decision-making processes. Neurotechnology can be defined as:

The assembly of methods and instruments that enable a direct connection of technical components with the nervous system. These technical components are electrodes, computers, or intelligent prostheses. They are meant to either record signals from the brain and 'translate' them into technical control commands or to manipulate brain activity by applying electrical or optical stimuli (Müller and Rotter 2017, 1).

For the purposes of this discussion, such mechanisms are treated as enabling conditions that underlie the theoretical arguments outlined earlier.

The premise can thus be formulated as follows: contemporary neurotechnological developments appear to allow third parties to interfere with a person's brain processes before those processes manifest as speech or action, thereby eroding the traditional boundary between the internal and external dimensions of the human mind—a conceptual shift that aligns with Derrida's notion of deconstruction. This interference can occur through at least two distinct pathways:

1. Access to thoughts and mental information. Neuroscientific technologies, such as brain scanners and brain-computer interface devices (BCI), hold the potential to access, interpret, and influence brain activity, thus raising concerns regarding mental privacy, or neuroprivacy. Advances in decoding brain activity are providing neuroscientists with powerful tools to enhance understanding of brain functions while testing theoretical predictions about the relationships between brain activity, cognition, and behavior (Loriette et al., 2022). BCIs, for example, enable individuals to control external devices such as computers or robotic prosthetics using neural signals. These devices record brain activity and translate it into digital commands (Yang et al., 2015). Companies such as Neuralink have even developed BCIs capable of capturing and processing information directly from the brain (Musk, 2019).

For instance, when a person contemplates a physical action, the neural signals captured by a BCI can potentially reveal not only the intention behind the action but also other thoughts that are not explicitly communicated. This opens the door to the interception or alteration of thoughts before an individual has full control over them. Moreover, advanced brain scanners like Functional Magnetic Resonance Imaging (fMRI) allow researchers to observe brain activity in real time. These technologies have been employed to decode thought patterns, identify emotions, and even detect intentions prior to their external expression. For example, in 2008, researchers demonstrated that they could predict with 60% accuracy the decisions individuals would make approximately seven seconds before the subjects themselves became aware of their choices (Soon et al., 2008).

More recently, in 2023, researchers at the University of Texas at Austin developed a non-invasive semantic decoder system that uses fMRI scans to translate brain activity into text. Participants in the study either listened to podcasts or imagined telling a story while their brain activity was recorded. The algorithm, trained on this data, successfully decoded the general content of their thoughts, though not with word-for-word precision. For example, when a participant listened to the phrase "I don't have my driver's license yet", their brain activity was decoded as "She has not even started to learn to drive yet". This illustrates the system's ability to capture the general meaning or "gist" of what was being thought or heard (Tang et al., 2023).

2. The access to thoughts and mental information described above can lead to profound cognitive and moral manipulation, significantly influencing individuals' perceptions and decisions. While digital platforms already utilize algorithms to personalize content and subtly shape user behavior, direct intervention in mental processes amplifies this potential, posing far-reaching ethical concerns. If third parties can access individuals' thoughts or emotions, they could exploit this information to influence choices—from product purchases to political decisions—thereby undermining personal autonomy. This phenomenon is commonly referred to as "indirect mind control". Below are examples illustrating how neurotechnological advancements may facilitate such manipulation:

    (a) Deep Brain Stimulation (DBS) for Emotional and Behavioral Control. DBS involves implanting electrodes in specific brain regions to modulate neural activity. Originally developed for medical purposes such as treating Parkinson's disease and depression, DBS has demonstrated the capacity to alter cognitive functions, including decision-making and emotional regulation. Studies reveal that stimulating certain brain areas can significantly influence an individual's mood, motivation, or risk tolerance during decision-making processes. These capabilities raise ethical concerns about the potential for external control over a person's emotions and behavior.
    (b) Memory Modification During Sleep Using TMR (Targeted Memory Reactivation). Research has shown that memories can be enhanced or weakened by reactivating them during specific phases of sleep through auditory stimuli. This process, known as Targeted Memory Reactivation (TMR), exposes participants to sounds associated with previously learned tasks while they sleep. Results indicate that TMR can strengthen or weaken memories, demonstrating how neurotechnology may influence memory consolidation. These findings raise ethical questions about the potential for manipulation

of memory content without an individual's informed consent (Rawson & Jackson, 2024).

(c) Brain-Computer Interfaces (BCI) for Mind Control. BCIs allow individuals to control external devices using brain activity alone. While these interfaces are frequently used for medical purposes, such as enabling paralyzed patients to operate prosthetic limbs, the same technology can also be applied to influence or manipulate cognitive processes. BCIs have been utilized to detect emotional states or cognitive intentions and relay this information to external systems, potentially enabling manipulation of decision-making or behavior in real time, often without the user's consent (Baselga-Garriga et al., 2022, 161).

(d) Transcranial Magnetic Stimulation (TMS) for Modifying Decision-Making. TMS uses magnetic fields to stimulate specific brain regions, temporarily influencing neural activity. This technique has been shown to affect cognitive processes, including decision-making and moral judgments. In experimental settings, researchers have used TMS to alter participants' choices in moral dilemmas, effectively demonstrating how cognitive functions can be externally manipulated through neurotechnology (Jeurissen et al., 2014).

(e) Emotion Recognition and Manipulation through Neuroimaging. Advanced neuroimaging techniques, combined with machine learning, can detect and decode emotional states by analyzing brain activity. Once emotional states are identified, external stimuli can be deployed to modify or influence these responses. For instance, neuroimaging can detect stress or anxiety levels, enabling systems to deliver targeted stimuli designed to induce calmness or regulate emotions in real time. This capability is particularly relevant for emotion self-regulation training in patients with major depressive disorder (Zotev et al., 2020).

(f) Optogenetics for Behavioral Control. Optogenetics involves genetically modifying neurons to make them responsive to light, allowing researchers to control neural activity using light stimulation. While currently limited to animal studies, this method has been used to induce aggressive behavior or diminish fear responses in mice by targeting specific brain circuits with optogenetic tools. These findings demonstrate how neural circuits responsible for complex behaviors and emotions can be manipulated externally, highlighting the potential for similar applications in human subjects (Mahadevia et al. 2021).

4 Law and the Deconstruction of the Inside and Outside of the Mind

The examples mentioned—though far from exhaustive—demonstrate how contemporary neurotechnology acts as a potential means of accessing third parties' mental states. Moreover, it initiates a deconstructive process that challenges the traditional dichotomy of "inside" and "outside" with respect to human actions and the intricate structure of decisions and thoughts that underpin them. This dimension, once thought to be private and inaccessible, now appears at least partially open to external intervention, raising profound ethical and philosophical concerns.

One might question the specificity of applying deconstruction to mental states and privacy, as well as the unique insights this perspective offers. To someone unfamiliar with Jacques Derrida's philosophy, the ability to access others' thoughts through neurotechnological methods without consent may simply suggest a new legal domain for protection—highlighting valid concerns about potential misuse. However, beyond this practical but somewhat narrow conclusion lies a deeper theoretical implication: the deconstruction of the inside/outside dichotomy fundamentally destabilizes a broader paradigm that underpins metaphysical and epistemological thought. This paradigm affirms the separation of empirical and transcendental, essence and existence, matter and form, being and nothingness, and mind and brain. In the context of mind and brain, this distinction traditionally elevated the mind as a spiritual principle, distinct from and superior to the corporeal brain.

The erosion of the boundary between signifier and signified, culminating in Derrida's concept of the "trace", further dismantles the notion of self-contained essences in both material objects and their mental correlates. Thought is no longer understood as the expression of an immaterial principle, but as a graphic sign—a mark whose meaning recedes into an open field of contingent interpretations. This conceptual shift transforms the act of thinking into a process analogous to reading or interpreting an "inner text," devoid of any definitive point of reference or pre-existing meaning. The signifier-thought becomes an empty structure, not oriented toward a fixed signified but instead toward an endless chain of other signifiers: "this totality of the signifier cannot be a totality, unless a totality constituted by the signified preexists it, supervises its inscriptions and its signs, and is independent of it in its ideality" (Derrida, 1997, 20).

Applying this deconstructive framework to consciousness suggests the emergence of a radically redefined concept of "mind". What was once regarded as the locus of ontological intimacy—be it the soul engaged in self-understanding, as described by Plato (1927, 132c-133c), or the *forum conscientiae*, representing true conscience beyond judgment, as articulated by Aquinas (2019, I-II, q.96, a.4)—now appears stripped of transcendental significance. It is instead reduced to a trace, readable within the physical substrate of the brain, akin to how a narrative is encoded in written symbols. According to this perspective, thoughts and mental states become mere epiphenomena of matter, with the brain and nervous system functioning as the substrate for these graphemic traces. This shift signifies a pivotal turn in the philosophical debate between dualism and monism, pushing the argument firmly toward materialistic monism.

Materialistic monism asserts that all phenomena, including consciousness, are reducible to material processes. In the 19th century, idealism dominated monistic thought, positing that minds and ideas constituted the only true reality, with material objects as mere derivatives. By the 20th century, however, idealism waned, giving way to materialism, which became the prevailing monistic theory, asserting that matter is the sole fundamental substance (Haugeland 1985, 38-9).

To dissolve the essential boundary between mind and brain—or between matter and spirit—it is necessary to propose a theory, such as Derrida's grammatology, that conceptualizes thought as inscribed and materially accessible. By framing thought as a grapheme, reducible to neural activity or material inscriptions, this perspective affirms the primacy of matter over any higher spiritual principle. In this framework, the mind is not an independent essence but a material trace, open to interpretation and devoid of an inherent, fixed identity. This deconstructive turn thus reshapes our understanding of consciousness, embedding it fully within the realm of material processes.

To fracture the essential boundary between mind and brain it is necessary to develop an argument supporting a general theory of writing—a "grammatology"—applied to thought. Such a framework would conceptualize thought as something written, rendered corporeally accessible to the "reader." In this perspective, thought can be captured and analyzed as brain activity, discernible through electrical signals or changes in blood flow. These patterns, recorded by technologies like EEG, fMRI, and BCI, can then be interpreted in real time using advanced AI algorithms capable of decoding thoughts, emotions, and intentions. This opens pathways to significant advancements in human-machine interfaces and critical applications in mental privacy and neuroethics.

The implicit conclusion is that mental activity aligns seamlessly with brain activity, framing thought as a signifier, readable in much the same way a footprint on the ground offers clues about its maker. Within this paradigm, thought is inscribed in the brain as letters are on paper, forming a universal mark—a general "writing" that encompasses what is thought, spoken, and even written. The brain becomes a unique kind of trace, a corporeal substrate, a biological grapheme that records these traces through neuronal operations and cognitive processes. Consequently, thought becomes a sign, a mark that—like any signifier in Derrida's framework—can be read and interpreted using appropriate tools. This inscription opens thought to deconstruction, dissolving the dichotomy between inside and outside as long as technologies exist that enable this legibility. However, these technologies do not create legibility but serve as its condition, allowing for an interplay where neither action nor speech is purely external, and neither thought nor consciousness remains entirely internal as traditionally conceived. Instead, both are recontextualized as contingent upon a boundary (limes) closed off by the overarching nature of writing.

The deconstruction of this mental boundary implies that consciousness could ultimately be reduced to cerebral "writing." In this light, everything human—body, cognition, feelings, and memories—becomes interpretable, deconstructible, and open to reconfiguration with each new reading of its records. Here, materialism finds full affirmation, requiring no principle beyond matter to account for the phenomena of the mind.

Some may argue that even within a dualist framework, any output produced by the "ghost in the machine" (Ryle, 1949) or other non-corporeal principles would still manifest in the brain, effectively reducing qualitative consciousness to signs. Yet this view remains problematic. If a spiritual principle exists, it produces thought as a distinct reality, with the spirit acting as its efficient cause, imbuing thought with a definitive nature (Aquinas *In Metaphysicam*, 1, V, lect. 2, n. 765, 2013). Within this framework, the brain serves merely as the organ through which thought operates, not as its cause. While the effect may reflect some aspect of its cause, it does not exhaust it. Thus, under dualism, brain "writing" cannot fully encapsulate the origin of thought, as the cause inherently surpasses the effect.

Neurotechnological advances enabling mind-reading profoundly alter the foundational assumptions of traditional law. Historically, legal frameworks have been confined to regulating and assessing individuals' outward behaviors, consciously refraining from encroaching on internal mental states. Access to another's mind has traditionally been limited to psychological constructs like the *Theory of Mind*, which posits those individuals—especially young children—rely on cognitive skills, not devices, to infer mental states and emotions (Wellman, 2020). The distinction between thought and action has always posed conceptual challenges: while thought precedes and influences action, suggesting a degree of independence, the relationship is less distinct when viewed through the lens of outward behaviors. Every intentional human action involves cognitive recognition and a voluntary element, which become integral to the action itself. Thus, legal evaluation has historically required tangible actions, as articulated by Thomasius in the 18th century, who asserted that the law's purpose is to maintain external social order by regulating outward conduct, without delving into matters of conscience or morality (Thomasius, 2008). Similarly, Kelsen emphasized that law requires "an external manifestation of human conduct" (Kelsen 1934/1967, 2), a position mirrored in Hart's analyses of coercive orders (Hart, 2012).

With the rapid advancement of neurotechnological innovations, the traditional boundary between internal mental states and external actions is increasingly blurred, necessitating a critical reexamination of legal principles and their application to emerging scenarios. Technologies such as brain-computer interfaces (BCIs) and advanced neuroimaging have introduced profound challenges, raising a host of legal and ethical risks, including:

1. Violation of mental privacy, or neuroprivacy. The ability to directly access individuals' thoughts and mental states introduces the potential for intimate information to be read, recorded, or used without consent. Governments or institutions could leverage technologies like fMRI to monitor the thoughts of employees or citizens, possibly identifying political dissidents and thereby infringing on the fundamental rights to privacy and freedom of thought.
2. Manipulation of decisions and free will. Techniques such as deep brain stimulation (DBS) and BCIs could be employed to modify a person's behavior, potentially compelling individuals to act against their will. These technologies may influence not only routine choices but also critical moral and ethical decisions.

3. Lack of informed consent. The use of neurotechnologies to access thoughts or manipulate decisions often occurs without sufficient informed consent, jeopardizing the principle of individual autonomy. For example, in clinical applications of transcranial magnetic stimulation (TMS), patients might not fully comprehend how the technology affects their emotions or decision-making processes. Additionally, neurotechnologies could covertly influence clinical decisions without the patient's awareness.
4. Neuro-surveillance and state abuse. Governments may deploy mental surveillance systems using neurotechnology to monitor thoughts, emotions, or intentions for social control or suppression. The implantation of BCIs or the use of brain-monitoring systems to detect subversive thoughts could pose severe risks to fundamental human rights, including freedom of expression and thought.
5. Neurodiscrimination. Access to brain activity could foster new forms of discrimination, where individuals are unfairly judged or penalized based on thoughts or emotions detected via neurotechnologies. For example, employers might use brain scans to assess concentration or stress levels, potentially discriminating against employees who fail to meet specific neurological standards, thus violating principles of equality and non-discrimination.
6. Criminal and civil liability. Neurotechnological manipulation of decisions raises critical questions about accountability. If a person's actions are influenced by external intervention, determining whether responsibility lies with the individual or the device operator becomes a significant legal challenge. Furthermore, non-consensual manipulation could lead to civil claims for emotional harm or violations of fundamental rights.
7. Ownership of brain data. The ownership of sensitive brain activity data remains undefined in legal contexts, creating opportunities for its exploitation. For instance, companies developing BCIs might collect and sell users' brain data to third parties, raising profound ethical concerns about the commercial use of such private information.
8. Manipulation of memories and emotions. Neurotechnologies like TMS or memory modification techniques during sleep can alter or erase memories, posing risks to personal identity and emotional well-being. Unauthorized use of these technologies could distort an individual's perception of reality, infringing on their right to mental integrity.
9. Military use of neurotechnology (neuroweapons). Neurotechnologies have the potential to be weaponized, altering the behavior of soldiers or adversaries in conflict situations. For instance, military applications could heighten aggression in troops or weaken the psychological defenses of opponents. Such actions might constitute war crimes and violate international humanitarian law.
10. Commercial exploitation. Advertising and marketing industries could exploit neuroscientific technologies to analyze and manipulate consumer thoughts and emotions. Neuromarketing strategies, which leverage brain data to personalize marketing messages, risk influencing consumer choices in ways that undermine autonomy and the right to make free and informed decisions.

5 Final Considerations

The above discussion leads to a twofold conclusion. First, (a) on a theoretical level, the deconstruction of the inside/outside dichotomy applied to the human mind suggests that all thoughts, even the most private and guarded, could eventually be accessed. This capability depends entirely on technological advancements. Such a premise implies that no mental state is, in principle, intrinsically protected from potential access by third parties equipped with advanced technical instruments. This conclusion could profoundly impact the concept of human dignity, as enshrined in legal texts, as well as personal freedom and autonomy. Their protection would shift from being universal and inalienable to a positive choice, thereby rendering them renounceable. Hypothetically, if an individual consents to allow another full access to and control over their psyche through neurotechnologies, no inherent moral or legal argument would prevent such an arrangement. Viewing the mind as a text composed of symbols (signifiers) that refer to each other in an indefinite chain suggests that various interpretations by different "readers" (including the thinker themselves) lack a fixed standard of accuracy. These interpretations remain valid even if they are contradictory or inconvenient for the individual or their community. This deconstructive reading further reinforces the identity of the brain and the mind, implying that artificial intelligence could eventually replicate the human mind in its entirety. Both systems, aside from their physical differences, could be considered autonomous formal systems of sign production. Here, "symbols" are used abstractly, without delving into the debate on whether classical computation involves "rule-governed symbol manipulation" while neural network computation is non-symbolic.

Second, (b) on a practical level, the dissolution of the internal/external boundary in relation to the mind necessitates an urgent legal response to protect citizens from the extensive personal and financial harm that these mental intervention technologies could cause. The rapid pace of scientific development demands the swift integration of "Neurorights"—individual rights designed to safeguard mental integrity—into national and international legal systems. This discussion could take place at both constitutional and statutory levels. Constitutional guarantees are generally understood as open catalogs; they inherently allow for the recognition of new rights and obligations emerging from their thematic scope. Open catalogs gain depth through the interpretation of existing norms. In this context, Neurorights could be incorporated into current fundamental guarantees, especially those related to personal data protection. This approach would mean that issues involving personal data, as they pertain to Neurorights, are already protected under the Constitution within the relevant fundamental guarantee (López-Silva & Madrid, 2021, 61). However, given the sheer volume, complexity, and rapid pace of innovation in neurotechnology, relying solely on constitutional or legal interpretation of existing regulations may prove inadequate. Explicitly incorporating general guarantees into constitutions, alongside detailed legislation addressing technically complex issues, would better serve the community by enabling legal practitioners to act with greater clarity and certainty in specific cases.

Thus, there are compelling grounds to argue that the use of emerging neurotechnologies to access and potentially manipulate thoughts can be understood as a deconstruction of the boundary between the internal and external aspects of the mind—if we assume that thoughts are "written" in the brain and are not derived from any qualitative principle beyond the realm of pure matter. However, if it were demonstrated that ideas originate from a source beyond the brain, no such deconstruction would occur; this scenario would instead represent an advancement in techniques for understanding the brain's functioning as an organ that channels intellectual activity (but does not generate it). In either case, it seems prudent to incorporate Neurorights into legal systems, as the erosion of mental privacy is already underway.

References

Aquinas, T. (2013). In Metaphysicam, Corpus Thomisticum https://www.corpusthomisticum.org/cmp05.html

Aquinas, T. (2019). Summa Thelogiae, Corpus Thomisticum, https://www.corpusthomisticum.org/sth2095.html

Aristotle. (2016). Metaphysics. Hackett Publishing Company, UK.

Baselga-Garriga, C., Rodríguez, P., Yuste, R. (2022). Neurorights. A Human Rights Solution to Ethical Issues of Neurotechnologies. In López-Silva, P., & Valera, L. (Eds.), *Protecting the Mind. Challenges in Law, Neuroprotection and Neurorights* (pp. 157–162). Springer

Derrida, J. (1987). The Truth in Painting (Original work published 1978), The University of Chicago Press.

Derrida, J. (1989). Edmund Husserl's Origin of Geometry: An Introduction (Original work published 1962). University of Nebraska Press.

Derrida, J. (1997). *Of Grammatology (Original work published 1967).* The Johns Hopkins University Press.

Derrida, J. (2011). The Voice and the Phenomenon. Introduction to the Problem of the Sign in Husserl's Phenomenology (Original work published 1967). Northwestern University Press.

Hart, H. L. A. (2012). *The concept of law.* Oxford University Press.

Jeurissen, D., Sack, A. T., Roebroeck, A., Russ, B., & Pascual-Leone, A. (2014). TMS affects moral judgment, showing the role of DLPFC and TPJ in cognitive and emotional processing. *Frontiers in Neuroscience, 8*(18), 1–9.

Kant, I. (2000). *Critique of the power of judgement (Original work published 1790).* Cambridge University Press.

López-Silva, P., & Madrid, R. (2021). On the convenience of including neurorights in the Constitution or in the law. *Revista Chilena De Derecho y Tecnología, 10*(1), 53–76.

Loriette, C., Amengual, J., Ben Hamed, S. (2022). Beyond the brain-computer interface: Decoding brain activity as a tool to understand neuronal mechanisms subtending cognition and behavior. *Frontiers in Neuroscience, 16*, 1–17

Mahadevia, D., Saha, R., Manganaro, A., Morgan, A. A., Dumitriu, D., Rayport, S., Chuhma, N., Ziolkowski-Blake, A., Ansorge, M. S. (2021). Dopamine promotes aggression in mice via ventral tegmental area to lateral septum projections. *Nature Communications, 12*:6796.

Musk, E. (2019). Neuralink. *An Integrated Brain-Machine Interface Platform with Thousands of Channels, Journal of Medical Internet Research, 21*(10), Article e16194.

Peñalver, P. (1995) Introducción a Derrida, J. La voz y el fenómeno, Pre-textos, Valencia.

Plato, A I. (1927). Harvard University Press (Loeb Classical Library).

Rawson, G., Jackson, M. L. (2024). Sleep and emotional memory: A review of current findings and application to a clinical population. *Current Sleep Medicine Reports, 10*:378–385S.

Ryle, G. (1949). *The concept of mind*. Hutchinson's University Library.
Saussure, F. (2013). *Course on general linguistics*. Bloomsbury Academics.
Soon, C., Brass, M., Heinze, H. J., et al. (2008). Unconscious determinants of free decisions in the human brain. *Nature Neuroscience, 11*, 543–545.
Tang, J., LeBel, A., Jain, S., et al. (2023). Semantic reconstruction of continuous language from non-invasive brain recordings. *Nature Neuroscience, 26*, 858–866.
Thomasius, C. (2008). Fundamenta iuris naturae et gentium ex sensu communi deducta. In Wang, B., Antony, J., Lurie, S., Brooks, P., Paller, K., Norman, K. (Eds.), (2019). Targeted memory reactivation during sleep elicits neural signals related to learning content. *Journal of Neuroscience 39*(34), 6728–6736.
Wellman, H. (2020). *Reading minds: How childhood teaches us to understand people.* Oxford University Press.
Yang, C., Virk, C. S., y Yang, H. (2015). Earable Sensors and Robots, Springer.
Yuste, R., Goering, S., Arcas, B., Agüera, Y., Bi, G., Carmena, J. M., Carter, A., Fins, J., Friesen, P., Gallant, J., Huggins, J. E., Illes, J., Kellmeyer, P., Klein, E., Mar-blestone, A., Mitchell, C., Parens, E., Pham, M., Rubel, A., Sadato, N., Sullivan, L., Teicher, M., Wasserman, D., Wexler, A., Whittaker, M., y Wolpaw, J. (2017). Fourethical priorities for neurotechnologies and AI. *Nature. 551*, 159–163.
Zotev, V., Mayeli, A., Misaki, M., Bodurka, J. (2020). Emotion self-regulation training in major depressive disorder using T simultaneous real-time fMRI and EEG neurofeedback. *NeuroImage: Clinical, 27*, 1–16.

Criminal Law, Prognosis of Criminal Behavior, Neuroscience and the Case of *Psychopathology*

Gustavo A. Arocena

Abstract In this article, the author analyzes the problem of predicting criminal behavior, as a central element of Argentine penitentiary law. To do this, the author describes clinical and statistical methods used by criminal courts, noting their virtues and disadvantages. Then, he postulates the convenience of resorting to the use of "composite" prediction protocols. Finally, the author focuses on the possible contributions of neuroscientific tests in terms of predicting violent, antisocial or criminal behavior.

Keywords Prediction of criminal behavior · Clinical prediction · Actuarial prediction · Neuroscience · Neuroprediction

1 Introduction

The Argentine legal system focuses on the probable future conduct of a person convicted of a criminal act in order to enable certain legal consequences linked to the execution of the prison sentence. For example, article 13 of the Penal Code prescribes that a person sentenced to a prison sentence who has served part of the sentence may be released—before it expires and under certain conditions—following a report from experts that make an individualized and favourable prediction of his social reintegration. In turn, article 53 of the same digest establishes that a person sentenced to a custodial sentence who has been sentenced to imprisonment for an indefinite period of time as an accessory to the last sentence may be granted conditional freedom after five years of serving the accessory confinement and provided that he has demonstrated attitudes that allow it to be plausibly assumed that he will not constitute a danger to society.

Meanwhile, Law 24.660, on the Execution of the Deprivation of Liberty Sentence, subordinates the granting of conditional freedom to "...the well-founded reports of

G. A. Arocena (✉)
Law School, Universidad Nacional de Córdoba, Córdoba, Argentina
e-mail: gustavo.alberto.arocena@unc.edu.ar

P. López-Silva (ed.), *Contextualizing Neuroprotection: Latin American Perspectives on the Impact of Neurotechnological Development in Life and Society*, Ethics of Science and Technology Assessment 54, https://doi.org/10.1007/978-3-031-96055-0_6

the technical-criminological agency, the Correctional Council of the establishment and the management of the penitentiary establishment that predict in an individualized manner…[the] social reintegration of the convicted person". Article 54, fifth paragraph, of the last set of regulations, states that the competent jurisdictional organ must deny the incorporation of the convicted person into the assisted freedom regime "…when it considers that discharge may constitute a serious risk for the convicted person, the victim or society". Finally, article 104 of this Argentine penitentiary law provides that the conceptual qualification of prisoners will serve as a basis for the application of the progressiveness of the penitentiary regime, the granting of temporary releases, semi-freedom, conditional freedom, assisted freedom, the commutation of sentence and the pardon, while its article 101 stipulates that the concept will be understood as the weighing of his personal evolution from which his greater or lesser possibility of adequate social reintegration can be deduced.

In the field of penitentiary doctrine, "social reintegration" is understood as the return of the convicted person to social life after serving all or part of his sentence to a custodial sentence, which will be "adequate" when the subject adjusts his external behavior to requirements of the law. At the same time, the danger to society or risk to it, to the convicted person himself or to the victim, to which the aforementioned legal provisions allude, refers to the probability that, upon regaining his freedom, the individual will engage in harmful behavior for peaceful coexistence, whether criminal in nature or even criminally atypical, but equally harmful to social peace.

As can be seen, the prognosis of the conduct of the convicted person when he regains his freedom becomes an element of capital importance for Criminal Law, so the protocols or instruments used to carry out that weighing are equally relevant. These are instruments that monitor the so-called "criminal risk factors", which are shown to be associated with a greater probability that an individual acts antisocial, violent, aggressive, and which include criminogenic variables (individual characteristics that increase the risk of recidivism), demographic, socioeconomic, and individual (e.g., intelligence, low level of frustration tolerance, etc.). The psychology of crime classifies them into static factors—which are historical and do not change (e.g., gender, criminal history, suffering from childhood abuse, specific brain injuries, etc.)—and dynamic factors—which, in principle, are changing and therefore provide the opportunity to intervene, modifying the future (e.g., impulsivity, drug use, social support, work, compliance with treatment, etc.)—(Tortora et al., 2020, 2).

There are various tools aimed at predicting likely future antisocial behavior. For the purposes of this chapter, and for simplicity, we will predominantly mention only two, namely: clinical predictions and actuarial predictions. We will also say that clinical protocols have had, in our institutional environment, a predominant use. Indeed, the jurisdictional organs resort to the expert opinion of psychiatrists, psychologists and social workers from specialized technical teams, who, prior to interviewing the convicted person and/or reading their criminological history, provide individual information about them and forecasts about their probable future behavior, so that the magistrates can resolve issues related to penitentiary execution, among which are the early releases of the convicted person. The thesis that we will defend is that there are good reasons to expand the use of methods for predicting antisocial human behavior, in order to incorporate statistical strategies together with clinical methods, and that neuroscience can make interesting contributions to the configuration of composite or mixed evaluation methods.

2 Clinical Prediction of Antisocial Behavior

As Gaudet et al. (2016) explain, the oldest clinical prediction methods were developed by mental health professionals—generally psychiatrists or psychologists—who would evaluate an individual to determine if he was suffering from a mental illness or other medical condition and if it was likely that he would be dangerous in the future (Gaudet et al., 2016, 506). These types of evaluations can be structured or unstructured. Structured tests are psychometric ones, which are based on rigid questions and objective answers to measure cognitive and emotional abilities. On the other hand, in unstructured instruments, which are the most used in our justice administrations, the interviewed has more freedom to raise issues and influence the course of the interview, while the psychiatrist or psychologist uses their professional judgment and experience to reach their conclusions about the individual's current and future behavior. Unstructured techniques are called "projective" techniques, and they rely on ambiguous or open stimuli to explore the personality and internal world of the person. The responses to these stimuli do not have a single correct interpretation. These tests are based on psychoanalytic theories and the psychodynamic model. The professional who uses them must have a high level of expertise in theory and technique. This class of instruments includes, for example, the following exams:

1. Rorschach Test, which consists of a psychological test that uses 10 sheets with symmetrical ink blots of ambiguous appearance, which the expert must present in a certain order to the person being evaluated, asking in each case "What could this be?", without adding any further indication. Among the indicators analyzed in this test, we find the number of answers, the frequency of the answers given in the general population or the predominance of certain contents.
2. Holtzman Inkblot Test, which is made up of 47 images of inkblots and, with the same dynamics as the Rorschach Test, is based on the scoring of different variables that arise from the individual's responses.
3. Murray's Thematic Apperception Test, which is made up of 31 sheets of which only 20 are applied to each individual, depending on their biological sex and age, so that, based on scenes related to topics such as family, fear, sex or violence, prepare a story that includes a past, a present and a future, with which the expert analyzes the needs and psychological pressures of the evaluated person.
4. The Human Figure Drawing Test, which, through this illustration, analyzes the intellectual maturity, evolutionary level and interpersonal relationships of the subject, focusing on their vision of themselves, others, their environment and their relationship with the world.

5. The Person in the Rain Test, which evaluates the self-image that the individual presents in an unfavorable condition such as rain. Through drawing, the inner world of the subject is revealed, his thoughts, his emotions and his defenses against stressful situations. The expert asks the interviewee to provide two types of response, namely: one graphic and the other verbal. The subject must make the drawing, and then tell a story that tells the situation that has been captured on paper.

Now, reliability and validity of these unstructured strategies is highly variable because the content and scope of the interview is always unique and because different physicians may place varying amounts of weight on different weighted factors. Additionally, clinicians may allow their emotions, prejudices, or loyalty to one party or the other to influence their judgments, while the goal is to avoid any evaluator bias that could have an effect on the results. In Gkotsi and Gasser's words:

> The main drawback of the clinical model is that it predominantly depends on clinical and subjective criteria, which often vary interindividually and intraindividually by being assessed by different psychologists and psychiatrists (…). The fact that each expert is based on their own intuition, experience, and knowledge can lead to different results in the assessments of an offender's level of responsibility and of the perpetrator's future probability of conducting further severe crimes. (Gkotsi & Gasser, 2016, 60)

3 Statistical or Actuarial Prediction of Antisocial Behavior

Specialized literature assures the superiority of these statistical methods with respect to clinical methods. These are instruments that attempt to evaluate individual risk using information derived, not from individual weights, but from data about groups with common traits. This strategy is, today, in common law, promoted as the most plausible based on the available evidence. Thus, for example, Poldrack et al. (2017) claim:

> Across nearly every domain tested, actuarial approaches have been shown to outperform clinical prediction (…). That is, in head-to-head contests between predictions that reflect the subjective judgments of human experts and predictions based on validated statistical models, the statistical models have nearly always won. (2)

It has been this circumstance, precisely, that has made actuarial prediction decisively make its way into the field of evaluation of probable future antisocial or criminal behavior. Experts highlight:

> Things have changed when it comes to prediction. This change has been of a technical nature, since a new scientific apparatus has been built that has partially abandoned the clinical diagnosis of dangerousness and has replaced it with the probabilistic estimation of the risk of violent and criminal acts: the so-called violence risk assessment, which is a technical procedure to estimate the probability of future occurrence of a behavior through the use of random variables or unidimensional predictions, such as unstructured clinical

> assessment, assessment based on structured clinical judgment, and actuarial assessment. (Ruiz Guarneros, 2020, 72)[1]

Nevertheless, and as we will see later, in recent decades there have been important transformations in issues of predicting people's behavior, which drive the incorporation of the contributions of neuroscience, and even computer science, through of the creation of brain-computer interfaces, to not only understand the structure and morphology of the human brain, but also investigate its functioning. This responds to the fact that the evolution of expert thinking in this field has even led to questioning, not only clinical methods, but also statistical ones, as a consequence of their limited predictive value. Hence, the evaluation of criminal risks based on algorithmic tools is postulated. Actuarial prediction has been consistently challenged by virtue of its statistical nature, which, due to its generality, conflicts with the judicial need to evaluate the probable criminal risk of a specific individual. At the same time, they are models that are based on relatively immutable factors that, therefore, hinder any perspective of change and evolution of the subject.

4 The Need to Complement the Use of Clinical Prediction Methods with Statistical Methods

We have seen that many experts assure the predictive predominance of statistical methods in relation to clinical ones. For this reason, we should note that, considering the state of the matter in Argentina—where clinical instruments are used almost exclusively-, the current nature of our criminal processes advises the complementation of clinical protocols with actuarial ones. Furthermore, the "integrated" use of different weighting instruments is presented as a plausible strategy based on the virtues and limitations of each tool. In this direction, Poldrack et al. (2017) state:

> In recent years, many behavioral instruments aimed at predicting crime have been published that are not adequately characterized by a simple clinical–actuarial dichotomy. Rather, the behavioral risk assessment process now exists on a continuum of rule-based structure, with completely unstructured ("clinical") assessment occupying one pole of the continuum, completely structured ("actuarial") assessment occupying the other pole, and several forms of partially structured assessment lying in between. (3)

Through this type of strategies, specific dimensions of human behavior and people's characters can be assessed, at the same time traits common to certain groups, all of which makes it possible to predict antisocial or violent behavior with a greater degree of precision, although very far from being infallible.

Demands for the improvement of prediction instruments based on the importance of the legal issues subordinated to their results, and even the demands inherent to a culture of social fear that has elevated public safety into a supra protected legal interest, have derived in a Criminal Law that thinks about the possibility of finding

[1] The original text, written in Spanish language, has been translated by us.

in neuroscientific verifications the necessary contributions for said improvement in the prediction of criminal risk. The limitations and deficits—already mentioned—inherent to clinical and statistical models have opened the doors to the possibility of superior neuroscientific contributions:

> Assessments of dangerousness …are typically individualized in the sense that they rely on the clinical assessment of psychiatrists (typically) who have interviewed the convicted person or read this person's medical file. Such assessments are controversial, because they are notoriously unreliable (…). This is where neuroprediction —e.g. neuroimaging of the brain to establish a frontal lobe disorder-enters the picture. (Lippert-Rasmussen, 2014, 125)

5 Neuroprediction of Criminal Behavior

Crime is, definitely, a complex and multi-causal phenomenon, which originates and is maintained by various circumstances, which include, among many others, the interaction of the subject with the context, the breaking of his social ties, their vulnerability and, in many cases, specific biological traits of the subject (e.g., low activity of the frontal lobe). Regarding the latter, neuroscience findings seem to have enough to say, thus contributing to the necessary improvement of methods for predicting generically antisocial and specifically criminal human behavior.

On the point, Tortora et al. (2020) emphasize:

> Behavioral traits can be correlated, sometimes strongly, with features of the human brain, and this raises new possibilities for predictive algorithms to be developed, allowing the prediction of dispositions of an individual. (3)

This correlation between certain characteristics of a person's brain and certain types of behavior has been confirmed, as we will see below, in cases of psychopathy, with respect to which high rates of criminal recidivism have been recorded. Well, research shows that the population group with this particular clinical condition shares many behavioral traits, such as, for example, behavioral disinhibition or lack of empathy, and that these characteristics seem to have common neurobiological bases.

According to structural and functional results from functional magnetic resonance imaging (fMRI), maladaptive behavior in psychopathy, including violence, may arise from dysfunctions within the corticolimbic and corticostriatal circuitry involved in affective arousal, emotion regulation, and decision-making based on values. For example, psychopathic individuals exhibit decreased gray matter volume in the amygdala and ventromedial prefrontal cortex, as well as reduced cortical thickness. Likewise, psychopaths show reduced recruitment of the amygdala and ventromedial prefrontal cortex during fear conditioning and moral decision-making, attenuated amygdala response during affective perspective taking, and reduced engagement of the ventromedial prefrontal cortex in responses to empathogenic stimuli and facial emotional stimuli. Reduced functional and structural connectivity between the amygdala and ventromedial prefrontal cortex has also been reported in psychopathies. In sum, there is empirical evidence that the socioemotional deficits of psychopaths arise from dysfunction of the corticolimbic circuit.

While corticolimbic dysfunction appears to play an especially prominent role in the interpersonal affectivity dimension of psychopathy, recent work highlights the importance of cortico-striatal dysfunction in the disorder's antisocial impulsivity symptoms. Closely related to all this, experts also report:

> One of the strongest and most widely studied risk factors for recidivism is impulsivity, or behavioral disinhibition, the persistent lack of restraint and consideration of consequences (3). Risk assessments, personality tests, and neuropsychological measures have been used to assess impulsivity and have demonstrated the ability to predict future antisocial behavior. (Aharoni et al., 2013, 6223)

Then, specifically, and as a reaffirmation of what we have expressed previously, add:

> The brain regions associated with impulse control have been well characterized. Consistent among these regions is the anterior cingulate cortex (ACC), a limbic region associated with error processing, conflict monitoring, response selection, and avoidance learning. (Aharoni et al., 2013, 6223)

With particular regard to violent antisocial behaviors, Poldrack et al. (2017), after warning about the difficulties of reconciling the static nature of laboratory evaluations with the dynamic nature of real-world violence, emphasize:

> [V]iolent behavior typically results from the interaction of trait-like vulnerabilities in capacities related to self-control and emotion regulation [6] with time-varying state factors (e.g., stress level and sleep deprivation) and punctate eliciting events (e.g., provocation). (5)

Growing interest in so-called "neuroprediction" is therefore due to the supposed biological proximity between brain-based measurements and the causal processes that produce violent behavior, which makes it reasonable to presume that assessments of brain activity could support better predictions. If neuroscientific experiments can demonstrate that criminal behavior has a neurobiological basis or correlation, this should intensify judicial interest in the possible application of neuroscience to criminal law. Neuroimaging techniques have great potential, although there are concerns about their application to complex real-world behaviors. Neurobiological measurements are not only more difficult to collect than behavioral, social and psychological data, but, as we have just noted, they can present difficulties in predicting the complex and multi-causal behavior of a specific person.

At this point, we can propose a definition—neither the only nor the last-of neuroprediction, characterizing it as the use of image data from the human brain to predict how a person will behave in the future, from the recognition that human thoughts and choices are based, at least partially, on underlying biological processes. In other words, we could assert that the neuroprediction of crime consists of the discovery of neurobiological markers of criminal recidivism. As an example of some of the verifications that neuroscientific studies have achieved, we can mention the following (Ruiz Guarneros, 2020, 73):

1. The amygdala serves to express subtle emotions such as love, affection, friendship, distrust and fear, and is very small in violent or very violent people.

2. The hypothalamus helps create basic emotions such as anger and terror, and positive states ranging from moderate pleasure to ecstasy.
3. The hippocampus helps the brain focus on sensory stimuli and the generation of emotions, linking these to memory, images and learning.
4. The anterior cingulate cortex (AAC)—as we mentioned in previous lines—is involved in problem solving, emotional self-control and error recognition, and its correct functioning enables adequate human socialization.
5. The prefrontal lobe is involved with the capacity for abstraction, intelligence, planning, inhibition of inappropriate behavior and regulation of emotions.

Now, based on everything we have noted, from an intuitive point of view I could maintain that the "problem"–*sit venia verbo*–with biological markers is the risk of incurring a kind of new criminological positivism that postulates biological determinism, according to which human behavior is defined by individual genes; in other words, by the hereditary load and the innate factor.

However, the conception—which we defend—of crime as a complex and, fundamentally, multi-causal phenomenon, places biological criminogenic factors as part of the many conditioning factors of criminal behavior; perhaps as an element of outstanding significance, but, in most cases, never isolated, but rather interacting with other factors that affect criminal behavior.

In this orientation, Poldrack et al. (2017) claim:

> Whereas cognitive neuroscience has taught us much about the neural bases of basic cognitive, emotional, and social functions, we still know almost nothing about how trait-like aspects of brain function, structure, chemistry, or connectivity interact with social context and other dynamic environmental factors to determine real-world behaviors, such as violence. What we can safely infer from recent work in other domains is that the causal structure of these behaviors will be highly complex and multiply determined. (4-5)

Of course, according to the arguments developed, ignoring the possible "biological conditioning" of probable future antisocial behavior would be as questionable as attributing to it an exclusive causal relevance. In the event that we understand that current neuroscientific discoveries do not yet allow conclusive future risk predictions to be made, it does not seem reasonable to dispense with them absolutely, and the future incorporation of "neuradata" in intelligence prediction models artificial opens up as an auspicious possibility.

As Nadelhoffer and Sinnott-Armstrong (2012) state:

> [R]ecent advances both in violence risk assessment and in the neuroscience of violence give us grounds for cautious optimism when it comes to the potential promise of neuroprediction. From the gathering data on the structural and functional neural deficits associated with psychopathy to the data on the epigenetic risks for violence associated with the MAOA gene [enzyme that breaks down the neurotransmitters norepinephrine, serotonin and dopamine, and that, in the event of dysfunctions in their action on certain areas of the brain, can alter the regulation of emotions and behavioral inhibition], researchers are starting to use multi-modal techniques to collect and analyze data in order to shed important new light on the neural correlates of violent behaviour. (634-635)

If, as recognized by important consensus in the expert field, criminal risk factors—especially those linked to violent behavior—can be classified as (a) dispositional or demographic, (b) historical, (c) clinical, and (d) contextual, there are good reasons to postulate that traditional methods of clinical prediction of the risk of probable criminal or antisocial behavior are insufficient. Eventually, such strategies will be suitable to address, although incompletely, dispositional risk factors—such as, for example, age, gender, socioeconomic status, cognitive functions or the neurological condition of the subject- or clinical –such as, *verbi gratia*, the existence of psychiatric or personality disorders. However, its effectiveness seems limited for the weighing of historical factors –such as, for example, criminal history or historical abuse of drugs or alcohol- or contextual factors—such as, *verbi gratia*, aspects related to the environment and the interaction of the individual with it-, in relation to which statistical predictions can provide information of special relevance to complement the clinical evaluation.

Such actuarial estimates seem to constitute a fertile field for the adequate reception of consolidated neurobiological evidence that leads to the improvement of prediction tools:

> Machine-learning techniques such as random forest [commonly used machine learning algorithm, copyrighted by Leo Breiman and Adele Cutler, which combines the output of multiple decision trees to arrive at a single output] have already been documented to improve the prediction of future charges of homicide or attempted homicide using traditionally available demographic and social variables. If neurocriminology can identify replicable biological risk factors that provide incremental knowledge over and above the traditional variables that are currently used in dangerousness assessments, this would further aid violence prediction. (Glen and Raine 2013, 59-60)

However, even in terms of clinical prediction, clinical risk assessment methods could find significant improvement from the contributions of neuroscientific evidence, if this consistently verifies the correlation between the functioning or structure of the brain of an individual person and certain kinds of criminal or antisocial behavior, or certain disorders, such as psychopathy, that are closely linked to such behaviors.

6 Final Remarks

It is not prudent to draw definitive conclusions regarding the topic we address, given the development, still in its infancy, of neuroscientific research. But, just as the so-called "brain preparation theory" has forced us criminal experts to rethink the foundations of a criminal charge that presupposes a free will whose existence has not been empirically proven, verifications through neuroimaging of the determinants or Biological conditions of violence and antisocial behavior, or the violent tendencies of human beings, require the review of prediction strategies for probable criminal behavior that, as happens in Argentina, are limited to the exclusive use of unreliable clinical methods. For now, we should think about complementing the latter, through the incorporation of statistical information, when our courts have it available from reliable sources. Now, if neurosciences achieve consistent verification of the neural correlations of violent, antisocial or criminal behavior, a more rational and humane criminal law will not be able to make its eyes cobwebs of such contributions from hard sciences.

References

Aharoni, E., et al. (2013). Neuroprediction of future rearrest. *PNAS, 110*(15), 6223–6228.

Gaudet, L. M., et al. (2016). Can neuroscience help predict future antisocial behavior? *Fordham Law Review, 85*, 503.

Gkotsi, G. M., & Gasser, J. (2016). Neuroscience in forensic psychiatry: From responsibility to dangerousness. Ethical and legal implications of using neuroscience for dangerousness assessments. *International Journal of Law and Psychiatry, 46* 58–67.

Glenn, A. L., & Raine, A. (2013). Neurocriminology: implications for the punishment, prediction and prevention of criminal behaviour. *Nature Reviews Neuroscience, 15*, 54–63.

Lippert-Rasmussen, K. (2014). Neuroprediction, truth-sensitivity, and the law. *The Journal of Ethics, 18*, 123–136.

Nadelhoffer, T., & Sinnott-Armstrong, W. (2012). Neurolaw and neuroprediction: Potential promises and perils. *Philosophy Compass, 7*(9), 631–642.

Poldrack, R. A., et al. (2017). Predicting violent behavior: What can neuroscience add?. *Trends in Cognitive Sciences, 22*, 111–123.

Ruiz Guarneros, A. I. (2020). "Neuroprevención": ¿una alternativa para analizar la reincidencia delictiva? In *Revista Mexicana de Ciencias Penales*, INACIPE, núm. 12, septiembre-diciembre.

Tortora, L. et al. (2020). Neuroprediction and A.I. in forensic psychiatry and criminal justice: A neurolaw perspective. *Frontiers in Psychology, 11*, 220.

Case Studies and Practical Lessons

Neurotechnologies and Inequality: Insights from Latin America

Nicolás Alejandro Serrano

Abstract In this chapter I argue that the peripherical socio-economic position of most Latin American countries give researchers from those countries a particular perspective on the potential consequences of emerging neurotechnologies, their regulation or lack thereof. Furthermore, I argue that this perspective can bolster international discourse by highlighting the importance of specific topics, therefore complementing mainstream analysis with new proposals based on regional insights. In order to exemplify this, I consider the issue of access to enhancement neurotechnology, its perceived interest in the LATAM region, and how that interest could provide original answers against the "rights inflation" objection to the relevant neurorights.

Keywords Neurorights · Neuroenhancement · Access · Rights inflation · Latin America

1 Introduction

During the 2022 international round table *"La regulación de la neurotecnología en Chile"*, a renowned neuroethicist posed an important question:

> Isn't it curious that a rights declaration is coming from a country that had a long history with a dictatorship? Is it in a way compensatory and reactive? You know, if you have been deprived of rights, that is something you really want to affirm (...). So I wondered if you could comment on why Chile and why Argentina, for that matter.

I would like to take this opportunity to address this topic and hold that it is precisely because of the history and "peripherical" status of Latin American countries, that members of their societies have a privileged perspective on some of the potential undesirable consequences of emerging neurotechnologies. Moreover, I would also like to argue that this perspective is not a detriment to LATAM-originated arguments

---

N. A. Serrano (✉)
Universidad de Buenos Aires, Buenos Aires, Argentina
e-mail: naserrano@filo.uba.ar

P. López-Silva (ed.), *Contextualizing Neuroprotection: Latin American Perspectives on the Impact of Neurotechnological Development in Life and Society*, Ethics of Science and Technology Assessment 54, https://doi.org/10.1007/978-3-031-96055-0_7

and proposals. Instead, it is a feature that can bolster international discussion by highlighting conceptual spots which, while also being a threat to socio-economic "central" countries, may be disproportionally overlooked by other researchers.

In order to show this, I'll start in Sect. 2 by briefly outlining some current advancements in neurotechnology and the ethical debates surrounding them. I'll then review, in Sect. 3, the most trending topics in the neurorights literature and compare them with prominent LATAM interests, most importantly inequity and access. Then, in Sect. 4, I'll present the "rights inflation objection" to lenca and Andorno (2017) and Yuste et al. (2021) neurorights proposals in order to show, in Sect. 5, how LATAM-based insights could contribute to the general international debate. In particular, I argue that Yuste, Genser and Herrmann's proposed *right to fair access to mental augmentation* stands in a privileged position to avoid the rights inflation objection. Most importantly, my objective here is not to offer definitive answers to these complex topics, nor to represent the entire "LATAM perspective" in a single book chapter. I merely aim to vindicate the value of this perspective by means of a concrete example. Ideally, this piece will help others to meaningfully engage with Latin American insights about neurorigths and neurotechnology.

2 The Neurotechnology and Neurorights Debate

It is becoming common knowledge that we are on the brink of a neurotechnological revolution. The accelerated and interwoven advancements of brain-computer-interfaces (BCI), artificial intelligence (IA) and both medical-oriented and consumer-oriented neurotechnologies have made what once were impossible goals into imminent (if not already tangible) realities. In the medical field, neurotechnologies offer new ways to approach what have been considered intractable conditions for centuries, and they promise to become more efficient and reliable in the foreseeable future. From well-known applications, such as restoring hearing through a cochlear implant (Peters et al., 2010; Wander & Rao, 2014; Zhang et al., 2020), to increasingly standard procedures, such as using deep-brain stimulation devices (DBS) to stop tremors in patients with Parkinson's disease (Benabid, 2003; Bucur & Papagno 2022). From enabling blind patients to see shapes and letters (Fernández, 2018; Lewis et al., 2015), to creating prothesis that give sensory feedback to amputees (Fisher et al., 2023). From allowing ALS patients to browse and communicate through the internet (Verbaarschot et al., 2021), to having a paraplegic person give the inaugural kickoff at the Brazil 2014 FIFA World Cup by means of a BCI-controlled exoskeleton (Nicolelis, 2014), along with a host of other benefits for quadriplegic patients in general (Lee et al., 2019). All these advancements pledge to improve the life of those affected by a plethora of different conditions. Moreover, neurotechnological applications go beyond the restoration of health to medical patients. Direct-to-consumer products offer BCIs aimed at neurostimulation, pleasure and gaming, among other non-medical uses. In sum, these developments constitute a trend that amounted to

33.2 billion dollars inverted in neurotech firms in 2021 alone (NeuroTech Analytics, 2022).

However, not all advancements are aimed towards health and pleasure. As it often happens with new technologies, military applications are also on the forefront of both governmental and private interests. In fact, the United States Defense Advance Research Project Agency (DARPA) is currently among the bigger spenders in neurotechnological research and investment (Kosal & Putney, 2022). Known DARPA projects include those aimed at developing "silent talk" (i.e. the use of BCI implanted devices to allow direct and silent mind-to-mind communication among soldiers), "telepresence" (i.e. remote vehicle control), a "cognitive technology threat warning system" (i.e. BCI-enhanced binoculars for improved perception and enhanced threat detection) and, more pressingly, research into the elimination of specific, targeted memories (Czech, 2021; see also Regalado, 2015; Kellmeyer, 2023). Yuste et al., (2021) also point towards future BCI potential for interrogation purposes as a way to directly access stored information, something that they consider could imply potential risks to national security. Military applications often overlap with a wider interest in *neuroenhancement* applications. That is, neurotechnologies that do not aim to restore lost or damaged capacities, but to improve those already existing beyond the current standards of healthy individuals. Some typical examples include the potential use of BCIs to increase memory consolidation and retrieval capacity (Burke et al., 2015) and general indices of attention (Rêgo et al. 2022; Prem et al., 2021).

Last, but not least important, all this research generates huge amounts of "neural data", either intentionally or as a byproduct. Chneiweiss (2023, 48) characterizes "neural data" as data relating to the functioning or structure of an individual human brain that includes unique information about their physiology, health, or mental states. As he points out, this data is considered a commodity in the consumer market and it is often repurposed for research in other domains, some of which may raise ethical concerns. For example, Cornejo Plaza (2023) points out how the recent surge in neuromarketing research (i.e. "the study of cerebral mechanisms likely to intervene in consumer behaviors") could affect our privacy (Nissenbaum, 1998; Véliz, 2021), cognitive freedom (Sententia, 2004), and mental integrity (Lavazza, 2018). Other domains, such as certain uses of affective computing (i.e. "the study and development of systems and devices that can recognize, interpret, process and simulate human affects") have spawned both legislation and ethical debates of their own (see Salles & Wajnerman Paz, 2024 for a discussion on these topics in social AIs).

In summary, while the potential benefits of the advancements in neurotechnology should not be understated, they carry with them an equal potential for both undesired and unintended consequences. As evidenced by the UNESCO's International Bioethics Committee report on *Ethical issues of neurotechnology* (IBC UNESCO, 2021), and the UNESCO's report on *The risks and challenges of neurotechnologies for human rights* (UNESCO, 2023a), current international debate tries to balance a recognition of the beneficial uses of emerging neurotechnology with an acknowledgment of its potential risks. The aim is to develop concrete suggestions for steering researchers, policymakers, and the public towards the best possible future, while avoiding both overly optimistic and pessimistic predictions and scenarios. It is in this context that the question for the need of neurorights emerged.

Neurorights can be initially understood as a new set of human rights explicitly aimed at protecting the mind, the brain and its data. More precisely, Ienca (2021, 6) defines them as "the ethical, legal, social, or natural principles of freedom or entitlement related to a person's cerebral and mental domain". The neurorights framework stems from a shared belief among several authors that current human rights either do not protect against the potential risks of emerging neurotechnologies or are not explicit enough to offer that protection in an effective way.[1] Currently, there seem to be two main neurorights proposals. One, made by Ienca and Andorno (2017, comprised of four new rights: cognitive liberty, mental privacy, mental integrity, and psychological continuity. The other, made by Yuste et al. (2017) and Yuste et al. (2021), proposes five new fundamental rights: mental privacy, personal identity, free will, fair access to mental augmentation, and protection from algorithmic bias. While there is some clear overlap between both proposals, one could argue that Yuste, Genser and Herrmann's seems to focus more on how neurotechnologies are going to be implemented and accessed in the near future. It could also be argued that because of this, Yuste, Genser and Herrmann's proposal had a resounding impact on Chile, the first country to include specific brain-related rights in its constitution (Rommelfanger et al., 2022; Cornejo Plaza, 2023).

3 Why Latin America?

As exemplified in the introduction, having Chile spearhead the legislation in neurorights may come as a surprise to some. And, given the relation most LATAM countries seem to have with neurotech development and research, this surprise could be justified. For example, when considering the "key countries which have advanced neurotechnology or artificial intelligence research programs", Yuste et al. (2021) mention the following nations: "United States, the UK, Canada, Australia, Russia, China, Japan, South Korea and applicable EU member states" (163). Importantly, not a single Latin American country has the kind of neurotech research programs needed to qualify in this list. Furthermore, with the possible exceptions of Brazil and Mexico, no LATAM country seems to share the economic status of those mentioned by Yuste, Genser and Herrmann (ECLAC 2023). Related to these quantifiable metrics, but in principle different from them, LATAM countries also seem to lack the capacity to "set the agenda" in what topics are relevant to neurotech research and development.

Let us refer to countries with the economic, political, and academic means to more easily and directly impact the course of neurotechnological research, development, implementation and international discourse as "central", and those lacking

[1] For an analysis of this "shared belief" as a non-neutral way of framing the neurorights debate, see Salles (2023).

such resources or capacities as "peripherical". An unsavory (but nonetheless legitimate) question could be: what meaningful contribution to the discussion could come from countries that are not directly involved in the relevant research and development processes? After all, cutting edge neurotechnology is mainly researched and developed in the central countries. Therefore, someone could argue that, for example, legislation (and their corresponding arguments) coming from a peripherical country is likely to be ill-advised, as it may be disconnected from the actual processes involved in the creation and implementation of neurotechnology.

However, I would like to argue that it is precisely because of their "peripherical" status, that stakeholders in LATAM countries can have a privileged view over certain aspects of the current neurorights debate.[2] In particular, given that these countries lack the economic and political means to directly impact neurotechnological research, development, and commercialization, their stakeholders are more prone to focus on the undesired consequences that (should they arise) their countries would be ill fitted to counteract. Crucially, it is not that these issues are not important by themselves from an ethical perspective, or that they aren't an issue for central countries. Rather, I believe that they disproportionally affect the more vulnerable countries and communities, making them a main concern for its members. And, while care should be taken to avoid turning this concern into exaggeration, care should also be taken to avoid the opposite phenomena, that is unintentionally minimizing potential undesirable consequences. In this regard, Latin American insights on issues of neurotechnological research, development and commercialization could contribute to spotting and addressing unattended or underrepresented topics in the international debate. And could, in turn, explain the pioneering attention that LATAM countries and legislation have paid to these topics.

As a matter of example, *access to neurotechnology* probably qualifies as a main topic of interest for LATAM stakeholders, as do the connected issues regarding the societal effect of widespread *neuroenhancement technology,* as well as the potential impact of its *military applications*. In each of these topics, LATAM countries often find themselves on the receiving end of developments that are mostly beyond their direct control, which makes their stakeholders take great concern with the less desired outcomes such developments could imply. For example, many LATAM scholars seem to be worried about the potential effects of uneven access to neuroenhancement, if this technology is to be generally adopted in the future. Some concerning effects include the potential segregation between a wealthy minority of enhanced individuals and unwealthy majority of unenhanced individuals, with its corresponding impact on the competitiveness of the local workforce. This can be seen, for example, in Cornejo Plaza (2023), Borbón et al. (2020), and Parlatino (2022, 2023) (although also see Salles, 2018), and explicitly addressed in the 8th principle of the UNESCO (2023b) report *Neurotechnologies and Human Rights in Latin America and the Caribbean*

[2] While this chapter focuses on the interests, concerns and initiatives from LATAM countries and researchers, it should be noted that many elements of this analysis could extend to other "peripheral" countries and, maybe, to the "global south" in general.

(redacted by the Argentinian jurist Roberto Andorno).[3] Another prominent example of concern regarding access is the recent Mexican proposal for a "General Law of Neurorights and Neurotechnology" (Senado Mexicano, 2024). Among other things, this initiative suggests changes in education laws in order to grant access to neurotechnologies relevant for neurodevelopment, stating that "The State shall promote and facilitate equitable access to neurotechnologies, eliminating economic and social barriers that impede fair access" (160, see also 37). Regarding military applications, most LATAM countries seem to face a paradoxical situation: while lack of access is concerning because of its defense implications, the troubled history that most LATAM countries have of military coups performed by their own military forces may raise an additional layer of concern about actually granting them access to certain forms of neurotechnology.

Importantly, while the general topic of "justice" has been generally acknowledged in the literature (Burwell et al., 2017), issues of equitable access and distributive justice appear to be among the *least* currently discussed in the international debate. Coin et al. (2020) made a comparative review of published academic discussion about the ethical implications of BCI technology during 2015–2020, aimed at improving and updating a previous comparative review made by Burnwell et al. (2017). Coin et al. (2020) identified the following topics to be the most trending, in order of most mentions: Autonomy [74%], Responsibility [71.4%], Humanity and Personhood [57.1%], Research Ethics and Informed consent [57.1%], User Safety [57.1%], Regulation [57.1%], Privacy and Security [42.8%], *Justice* [28.6%], Stigma and Normality [28.6%], Societal Implications [28.6%], *Military Use and Enhancement* [14.3%]. Crucially, the notion of "justice" present in this list doesn't reflect issues of *distributive* justice or unequitable access. As the authors point out:

> Justice was also mentioned infrequently [28.6%, n=2], with the main idea being inequality and injustice *within the research.* These discussions often related back to the aforementioned questions of when the trials would end and if the participants would get to subsequently keep the BCI technologies (Coin et al. 2020, 7, my emphasis).

Not only is the notion of "access" not mentioned among Coin et al. (2020)'s trending topics, it could be argued that some of the most frequently mentioned topics (e.g. "autonomy", "responsibility", "humanity and personhood", "user safety", and even "justice") gain part of their current relevance from a *presupposition* of general access to neurotechnologies. This should come as no surprise if one believes that the common themes identified by Burnwell et al. (2017) are somewhat correlated to the more pressing concerns of "central" countries (a position that could be bolstered by pointing out that their study only included papers written in English, therefore likely excluding debates in, for example, LATAM's local academic fora). The economic

---

[3] Such principle calls for caution regarding neuroenhancement technologies, warning that: "The use of such devices without therapeutic reasons could be socially problematic, as it could generate an unfair advantage for neuroenhanced individuals over others in various areas of social life, such as access to jobs or in the evaluation of school or university performance. In this sense, we should prevent the emergence of a social gap between people who have decided to increase their cognitive abilities and those who could not or chose not to do so". (UNESCO 2023b, 36).

status of central countries makes general access to neurotechnology more likely. And, even acknowledging that access inequality could still be an issue for these countries, it could be argued that their advanced research programs give them the tools to take measures against such inequility by locally developing cheaper neurotech alternatives.

However, peripherical countries like those in LATAM realistically face the threat of the opposite scenario, i.e. an *un*likely general access to neurotechnology, while also lacking the means to develop accessible and competitive neurotechnologies to counteract it. Among other things, this would imply the mentioned competitive disadvantage for LATAM's workforce in general, which would in turn likely exacerbate its vulnerable socio-economic position. Because of this, LATAM's countries are at the mercy of the developments in central countries (both for better and for worse). Crucially, this aligns with a historical trend of enduring different kinds of inequity on a daily basis. A trend that, in my opinion, has generally resulted in a disposition to approach most issues from an economic, cultural and class perspective. This, I would argue, makes the issue of access particularly salient for LATAM stakeholders. And, therefore, could make those who focus on these topics in LATAM be more aware of some issues that might be overlooked in other places.

Fortunately, LATAM's perspective doesn't end in the mere awareness of potential issues: it also encompasses a characteristic drive to create inventive solutions to pressing problems. Given the great importance that inequity issues have for LATAM societies, let us consider one of the proposed neurorights that directly focuses on these concerns: the right to fair access to mental augmentation (Yuste et al., 2021). On the remaining Sections I will analyze this right in the context of the rights inflation objection and try to offer some preliminary ideas towards a defense against it.

4 The Right Inflation Objection

As previously mentioned, neurorights can be initially understood as human rights explicitly aimed at protecting the mind, the brain and its data. However, beside this first characterization, there is considerable debate about how to better understand them, how necessary they actually are, and (if indeed necessary) how to best legally articulate them. In this respect, one matter of debate is whether neurorights should be considered as completely new additions to the list of human rights (e.g. Ienca & Andorno 2017; Yuste et al., 2021), as reconceptualizations of existing human rights (e.g. Muñoz & Marinaro, 2022, 2023), or even as merely new interpretations on already existing and unmodified human rights (e.g. Borbón et al. 2020; Fins 2022; Hertz, 2023).

One of the reasons behind this debate, and especially behind the "mere interpretation" stance, is a concern about the risk of *human rights inflation.* The main idea is that not everything that is morally desirable should be elevated to the status of a human right. As Hertz (2023, 4) points out "a large number of human rights leads to high compliance and monitoring burden and may compromise the effective protection of those human rights that are truly fundamental". This would lead to the paradoxical result of reducing protection by increasing rights. Furthermore, Ienca (2021, 9) also points out that "the unjustified proliferation of human rights is problematic because it may spread skepticism about all human rights, as it dilutes them to mere moral desiderata or purely rhetorical claims". These kinds of worries have lead authors like Borbón, Borbón and Pinzón (2020) to doubt the efficacy of including new neurorights, even while acknowledging the potential societal risks of emerging neurotechnologies (e.g. discrimination due to the lack of, and resulting pressure to adopt, neuroenhancement technologies, 153).

While there are multiple neurorights proposals, there are two main initiatives that seem to be the starting point of most discussions on this topic. On the one hand, there is Ienca and Andorno (2017)'s proposal to include four new neurorights. First, the *right to cognitive liberty*, strongly inspired in Bublitz (2013) and aimed at ensuring both the right of individuals to use neurotechnologies, as well as protecting them from coercive and unconsented use of such technologies. Second, the *right to mental privacy*, "which aims to protect any bit or set of brain information about an individual recorded by a neurodevice and shared across the digital ecosystem" (Ienca & Andorno 2017, 15). Third, the *right to mental integrity*, which offers "protection from potential neurotechnology-enabled interventions involving the unauthorized alteration of a person's neural computation" (18). And fourth, the *right to psychological continuity*, which "tends to preserve personal identity and the coherence of the individual's behavior from unconsented modification by third parties" (21).

On the other hand, there is Yuste et al. (2021)'s proposal of five neurorights. First, the *right to identity*, understood as "the ability to control one's physical and mental integrity" (160). Second, the *right to agency*, which encompasses "the freedom of thought and free will to choose one's own actions" (160). Third, the *right mental privacy*, which includes "the ability to keep thoughts protected against disclosure" (160). Fourth the *right to fair access to mental augmentation*, which seeks to "ensure that the benefits of improvements to sensory and mental capacity through neurotechnology are distributed justly in the population" (160-161). And finally, fifth, the *right to protection from algorithmic bias*, that is "the ability to ensure that technologies do not insert prejudices" (161).

Proponents of the rights inflation objection generally state that, while these proposals may be well intentioned and aimed at relevant and pressing issues, the protection they seek to offer is already conveyed by pre-existing human rights. For example, both proposals' rights to mental privacy could potentially be covered by article 12 of the Universal Declaration of Human Rights (UDHR), i.e. Right to Privacy. On the other hand, article 19 of the UDHR, i.e. Freedom of Expression,[4] could be interpreted to cover the intended purposes of Ienca and Andorno's right to cognitive liberty and Yuste, Genser and Herrmann's right to agency. Furthermore,

[4] Which reads "everyone has the right to freedom of opinion and expression; this right includes freedom to hold opinions without interference and to seek, receive and impart information and ideas through any media and regardless of frontiers".

both proposals' rights to mental integrity could fall under the umbrella of international human rights treaties, such as Article 3 of the Charter of Fundamental Rights of the European Union (i.e. Right to Mental Health). Lastly, insofar as it targets the insertion of prejudice, the right to protection from algorithmic bias could be cover by Article 2 of the UDHR (i.e. freedom from discrimination), while it could be argued that article 5 (i.e. Freedom from Torture) protects against unconsented threats to psychological continuity.[5]

5 The Special Case of Fair Access to Mental Augmentation

The previous quick overview isn't necessarily aimed at endorsing the right's inflation objection, but rather at pointing out what seems to be a clear outlier in this strategy: the *right to fair access to mental augmentation* (henceforth FAMA). As a general strategy I would like to argue that, while other neurorights deal with the particular effects that could be caused by neurotechnologies, FAMA focuses on *neurotechnologies themselves* as concrete enablers for certain effects. Therefore, it would be irrelevant if the *effects* of neuroenhancement technology could be protected by other rights. What this right is intended to grant is the access to the concrete medium itself (i.e. neuroenhancement technology), and not its effects. In this regard, I consider it similar to other high-profile proposals for "concrete enablers" focused rights, such as the right to universal access to the internet. There have been at least five resolutions "On the promotion, protection and enjoyment of human rights" by the UN Human Rights Council (2012, 2014, 2016, 2018, 2021) which partly aimed at enshrining Universal Internet Access as either a right of its own or as constitutive part of Article 19. However, this last strategy seems to be insufficient, as the UN Secretary-General annual report of 2021 explicitly called for achieving "universal access to the internet by 2030 *as a basic human right*" (United Nations, 2021, 4, my emphasis) and, more recently, Volker Türk, United Nations High Commissioner for Human Rights has been recorded stating that "It may be time to reinforce universal access to the Internet *as a human right*, and *not just a privilege*" (OHCHR 2023, my emphasis).

This similarity between FAMA and Universal Access to the Internet, contrary to what happens with other neurorights similarities to existing human rights, doesn't seem to undermine FAMA but to strengthen it. This is because, while it is quite undisputable that Universal Access to Internet does not deal with neurotechnology, it could be argued that (in a future where neuroenhancement is as necessary for human development and flourishing as the internet is today) there would be no clear reasons to accept one but not the other. In this regard, and adding to the mentioned UN resolutions, Sample et al. (2019) calls into attention how widespread use of certain technologies has turned them into a precondition for the realization of personhood. Therefore, if there seems to be actual steps being taken to enshrine Universal Access

[5] At least so long as such unconsented interventions could be considered as "cruel, inhuman or degrading treatment".

to the Internet as a human right because of the crucial role it now occupies in our economic and social life, the same reasons could be considered sufficient to eventually grant FAMA the same status (at least if it is assumed that neuroenhancement will likely have the societal and economic impact often predicted).

To this end, it is important to note that the objection of rights inflation seems to combine at least three complementary, but in principle different, strategies. First, as noted by Hertz (2023), the objection points toward the practical difficulties caused by the progressive inclusion of more rights. As mentioned before, these include a reduction of the cultural and societal impact of human rights and their legal applicability. However, while these are all valid considerations, they are also the type of concerns that should clearly be analyzed on a case-by-case basis. So, it could be argued that what is at stake is a risk/benefit analysis, where the ethical objective is to carry on the action that would produce the most benefits while having the least risks. Such analysis will naturally vary from proposed right to proposed right. So this side of the inflation objection doesn't, by itself, rule out any particular right. Rather, it calls for careful consideration of the importance and benefit of each proposed right in light of the risk of lowering the social impact and legal applicability of existing rights. But again, such analysis needs to be historically situated and shaped by cultural considerations. To illustrate, the risks of having an unwealthy and unenhanced workforce competing with a wealthy and enhanced workforce, and the socio-economic impact (both regionally and globally) of such a divide may have more resonance in the Latin American context because of its specific socio-economic position. It could be argued that the "digital divide" (i.e. the unequal access to digital technologies that prompted the discussions for Universal Access to the Internet) would pale in comparison to the "enhancement divide" such world would bring.

Second, as noted by Ienca (2021), there is the argument that rights inflation threatens human rights on a conceptual level. The main idea here is that human rights should not be equated with "anything ethically desirable", but rather with those basic and inalienable rights that are characteristic of a human being. However, it is important to note here that the mere notion of neuroenhancement technology is already having the effect of questioning and redefining what is to be considered as "essentially human". This has lead authors like Zehr (2015) to even consider the possibility of a potential transformation of our species, akin to the evolution into a "homo sapiens technologicus" (but see also Müller & Rotter, 2017; Sample et al., 2019; Coin et al 2020 for similar, although less extreme, positions). Even if one rejects Zehr (2015)'s position, it is enough to accept that the ongoing development of these technologies will likely force a revision on what we understand as essentially human, to reject an *a priori* dismissal of their potential relevance as something to be considered inalienable and characteristic of future human beings.

Third, and most importantly, the rights inflation objection can take the more direct route of stating the non-necessity of new rights by showing that their desired outcomes already obtain from previously existing human rights. In this regard, it was previously mentioned how some authors consider that most of the neurorights proposed by Ienca and Andorno (2017) and Yuste et al. (2021) are already covered by existing rights, either in the Universal Declaration of Human Rights or in the more general umbrella of international human rights treaties. However, precisely because FAMA focuses on the concrete enabler (i.e. enhancement neurotechnology) for certain effects, there isn't any *prima faccie* example of an already existing right that covers its intended target.

It could be objected that, under the anticipated future scenario that would make FAMA necessary, its intended target could fall under a specific interpretation of the Article 25 of the Universal Declaration of Human Rights, i.e. the Right to Adequate Standard of Living. After all, the whole argument for FAMA could be summarized as stating that unequal access to neuroenhancement would (in the predicted future) lead to an inadequate standard of living for those left unwillingly unenhanced. However, under a more thorough inspection this doesn't seem to be the case. This is because Article 25 points toward certain specific domains of applicability, as it states that:

> Everyone has the right to a standard of living adequate for the health and well-being of himself and of his family, including food, clothing, housing and medical care and necessary social services, and the right to security in the event of unemployment, sickness, disability, widowhood, old age or other lack of livelihood in circumstances beyond his control.

Neuroenhancement doesn't seem to fall under the general enablers of well-being, such as food, clothing or housing. Neither should it be considered as part of medical care, since the whole difference between neuroenhancement and therapeutic neurotechnologies is that the latter aim to restore compromised capacities to a pre-damage standard, while neuroenhancement technologies seek to *increase* capacities beyond their natural limits. On the other hand, "social services" are generally understood as government programs aimed at providing education, housing assistance and access to medical care, none of which seem to particularly affect access to neuroenhancement. Lastly, while FAMA advocates would argue that eventual lack of access to neuroenhancement technologies would likely result in unemployment, the kind of "security in event of unemployment" alluded to in Article 25 aims toward some form of *post hoc* unemployment insurance (and not to fair access to the tools that could *prevent* unemployment in the first place).

In summary, it could be argued that the right to fair access to mental augmentation is in a privileged position compared to other proposed neurorights when dealing with the rights inflation objection. Its risk/benefit profile seems adequate when considering the possibility of an "enhancement divide" akin to the current "digital divide". It can't be dismissed *a priori* as dealing with something uncharacteristic of (future) human beings, since the mere existence of neuroenhacenement technologies promises to modify how we understand what it is to be human. And, lastly, its targeted protections don't seem to be covered by already existing human rights. Being this so, FAMA and the issue of fair access to neurotechnology in general should receive far more attention than what Coin et al. (2020)'s review seems to show. A problem that is particularly concerning for those of us living in Latin America and other peripherical regions.

As a final consideration, some may worry that much of the neurorights debate isn't sufficiently grounded on real, scientifically informed, developments and possibilities, and that it can even fall prey to concerns merely inspired in science fiction scenarios. Since widely commercially available neuroenhancement is not yet a reality, and its actual impact on the workforce remains unrealized, it could be argued that this criticism applies to the position defended here. Ultimately, only time will tell if it actually does or not. However, I would like to conclude by offering three preliminary answers to these kinds of concerns. First, in this piece we have only considered the potential impact of real, actually in development, neurotechnologies. In light of this, I would argue that memory and attention enhancement, BCI access to internet data, or enhanced physical attributes through BCI controlled prosthetics don't qualify as fantasies ungrounded in reality. On the contrary, the expressed concerns are grounded in the potential effects that currently in development (if not already patented) neurotechnologies could have on society, economy, and the workforce in particular. Second, the predicted effects of unequitable access to neuroenhancement are also grounded in the already observed effects of unequitable access to other types of technology, such as the "digital divide". Those unable to access information and communication technologies (such as the internet) face verifiable and undesirable consequences at an educational, societal, and economic level, which has led to the creation of several funds and legislations aimed at mitigating them.[6] There is no reason to assume unequitable access to neuroenhancement wouldn't follow the same trend; while adding on top the way it directly affects the capacities of individuals themselves, and the additional effects that that could have on the perception of those left without access. And third, while it cannot be denied that a scenario of wide (but uneven) use of neuroenhancement technology is thus far speculative, its possibility is also grounded on verifiable trends in human history: access to literacy, combustion vehicles, computers, internet and cellphones were all once seen as luxury goods and/or used for few and very specialized types of labor. Nowadays, access to most of them is simultaneously uneven and a precondition for nearly all jobs. While there have been proactive calls to prevent the emergence of this kind of scenario with the nascent neurotechnologies before, what the LATAM perspective adds on top of this is a contextually situated awareness of the particular social, economic and cultural impact that such scenario would have, not only at the individual level, but for the region as a whole.

6 Conclusions

In this chapter, I've argued that LATAM's peripherical position, in socio-economic status and neurotechnological development, gives its researchers a particular insight into the potential negative effects of emerging neurotechnologies. While this isn't meant to imply that authors from other regions may not share these kinds of worries, it is meant to imply that authors from this region tend to have these matters at the forefront of their concerns. Furthermore, this also means that authors from this region

[6] For a recent example, in 2021 the Biden administration approved the bipartisan "Infrastructure Investment and Jobs Act" legislation, which allocated a 65 billion budget to increase connectivity in order to reduce the digital divide.

could offer new and interesting views on topics that could be underrepresented in the general literature. I tried to exemplify both points by analyzing the proposed "right to fair access to mental augmentation", and how it could be considered to avoid the rights inflation objection. Ultimately, it is important to stress that this chapter isn't meant to give some form of definitive account of the "LATAM perspective" on these complex issues. Rather, my intention is simply to encourage the consideration of (and discussion about) local and regional perspectives on these pressing debates of our time.

References

NeuroTech Analytics. (2022). Global Neurotech Industry Investment Digest 2021. Accessed at analytics.neurotech.com/neurotech-investmentdigest. pdf. (30 September 2022).

Benabid, A. L. (2003). Deep brain stimulation for Parkinson's disease. *Current Opinion in Neurobiology*. https://doi.org/10.1016/j.conb.2003.11.001

Borbón, D., Pinzón, J., & Borbón Rodríguez, L. (2020). Análisis crítico de los NeuroDerechos Humanos al libre albedrío y al acceso equitativo a tecnologías de mejora, *Revista Electrónica de Derecho y Ciencia, 6*. 135–161. https://doi.org/10.12795/IETSCIENTIA.2020.i02.10

Bublitz J-C. (2013). My mind is mine!? Cognitive liberty as a legal concept. In: Hildt, E., Franke, AG., (Eds.), *Cognitive enhancement. An interdisciplinary perspective* (pp. 233–64). Springer, Dordrecht.

Bucur, M., & Papagno, C. (2022). Deep brain stimulation in Parkinson disease: A meta-analysis of the long-term neuropsychological outcomes. *Neuropsychology Review*. https://doi.org/10.1007/s11065-022-09540-9

Burke, J. F., Merkow, M., Jacobs, J., Kahana, M. J., & Zaghloul, K. (2015). Brain computer interface to enhance episodic memory in human participants. *Frontiers in Human Neuroscience, 8*, 1055.

Burwell, S., Sample, M., & Racine, R. (2017). Ethical aspects of brain computer interfaces: a scoping review. *BMC Medical Ethics*. https://doi.org/10.1186/s12910-017-0220-y

Chneiweiss, H. (2023). Ethics issues and global governance of neurotechnology. In *The risks and challenges of neurotechnologies for human rights* (pp. 48–51). UNESCO.

Coin, A., Mulder, M., & Dubljević, V. (2020). Ethical aspects of BCI technology: What is the state of the art? *Philosophies, 5*(4), 31. https://doi.org/10.3390/philosophies5040031

Cornejo-Plaza, I. (2023). Chilean neurorights legislation and its rele-vance for mental health: Criticisms and outlook. *Salud Mental, 46*(5), 269–274.

Czech, A. (2021). Brain-Computer interface use to control military weapons and tools. In: *Advances in intelligent systems and computing*, (pp. 196–204). https://doi.org/10.1007/978-3-030-72254-8_20

ECLAC (UN's *Economic Comission for Latin America and the Caribbean*). (2023). Preliminary Overview of the Economies of Latin America and the Caribbean 2023, ECLAC, 2023.

Fernández, E. (2018) Development of visual Neuroprostheses: Trends and challenges. *Bioelectronic Medicine*. https://doi.org/10.1186/s42234-018-0013-8

Fins, J. J. (2022). The unintended consequences of chile's neurorights constitutional reform: moving beyond negative rights to capabilities. *Neuroethics, 15*(3). https://doi.org/10.1007/s12152-022-09504-z

Fisher, L. E., Gaunt, R. A., & Huang, H. (2023). Sensory restoration for improved motor control of prostheses. *Current Opinion in Biomedical Engineering*. https://doi.org/10.1016/j.cobme.2023.100498

Hertz, N. (2023). Neurorights—do we need new human rights? A Reconsideration of the Right to Freedom of Thought. *Neuroethics, 16*, 5 (2023). https://doi.org/10.1007/s12152-022-09511-0

IBC (International Bioethics Committee of) UNESCO. (2021). *Report of the International Bioethics Committee (IBC) on the ethical issues of neurotechnology*. Available at https://unesdoc.unesco.org/ark:/48223/pf0000378724

Kellmeyer, P. (2023). Neurotechnology and fundamental rights: conceptual and ethical foundations. In *The risks and challenges of neurotechnologies for human rights* (pp. 39–44). UNESCO.

Kosal, M. E., & Putney, J. (2022). Neurotechnology and international security. *Politics and the Life Sciences*. https://doi.org/10.1017/pls.2022.2

Lavazza, A. (2018). Freedom of thought and mental integrity: The moral requirements for any neural prosthesis. *Frontiers in Neuroscience, 12*, 82. https://doi.org/10.3389/fnins.2018.00082

Lee, M. B., Kramer, D. R., Peng, T., Barbaro, M. F., Liu, C. Y., Kellis, S., & Lee, B. (2019). Brain-computer interfaces in quadriplegic patients. *Neurosurgery Clinics of North America, 30*(2), 275–281. https://doi.org/10.1016/j.nec.2018.12.009

lenca, M. (2021). On Neurorights. *Frontiers in Human Neuroscience*, 15. https://doi.org/10.3389/fnhum.2021.701258

lenca, M., & Andorno, R. (2017). Towards new human rights in the age of neuroscience and neurotechnology. *Life Sciences Society and Policy, 13*(1). https://doi.org/10.1186/s40504-017-0050-1

Lewis, P. M., Ackland, H. M., Lowery, A. J., & Rosenfeld, J. V. (2015). Restoration of vision in blind individuals using bionic devices: A review with a focus on cortical visual prostheses. *Brain Research, 1595*, 51–73.

Mexicano, S. (2024). "Iniciativa con proyecto de decreto por el que se expide la ley general de neuroderechos y neurotecnologías y se reforman y adicionan diversas disposiciones en materia de neuroderechos y neurotecnologías", propossed by Senator Alejandra Lagunes Soto Ruíz on July 17th, 2024, Mexico.

Müller, O., & Rotter, S. (2017). Neurotechnology: Current developments and ethical issues. *Frontiers in Systems Neuroscience, 2017*(11), 93.

Muñoz, J. M., & Marinaro, J. Á. (2023). Neurorights as reconceptualized human rights. *Frontiers in Political Science, 5*, 1322922. https://doi.org/10.3389/fpos.2023.1322922

Muñoz, J. M., & Marinaro, J. Á. (2022). Algorithmic biases: Caring about teens' neurorights. *AI & Society, 39*(2), 809–810. https://doi.org/10.1007/s00146-022-01516-w

Nicolelis, M. A. (2014). Brain-to-brain interfaces: When reality meets science fiction. *Cerebrum, 2014*, 13. PMID: 26034520, PMCID: PMC4445586.

Nissenbaum, H. (1998). Protecting privacy in an information age: The problem of privacy in public. *Law and Philosophy, 17*, 559–596. https://doi.org/10.1023/A:1006184504201

OHCHR (Office of the High Commissioner for Human Rights), 2023. "It May be Time to Reinforce Universal Access to the Internet as a Human Right, Not Just a Privilege, High Commissioner tells Human Rights Council", 2023/03/10. https://shar.es/agq1Ao

Parlatino (2022). Declaración con recomendaciones sobre la necesidad de introducción de los neuro derechos en las legislaciones de los Congresos de este PARLATINO. Available at: https://parlatino.org/wp-content/uploads/2017/09/declaracion-neuroderechos.pdf

Parlatino. (2023). Ley Modelo de Neuroderechos para América Latina y El Caribe. Available at: https://parlatino.org/wp-content/uploads/2017/09/leym-neuroderechos-7-3-2023.pdf

Peters, B. R., Wyss, J., & Manrique, M. (2010). Worldwide trends in bilateral cochlear implantation†. *The Laryngoscope, 120*, S17–S44. https://doi.org/10.1002/lary.20859

Prem, A., Mohanraj, K., & Samuel, A. R. (2021). Brain computer interface (BCI) on attention: a scoping review. *Journal of Experimental Biology and Agricultural Sciences, 9*. https://doi.org/10.18006/2021.9(Spl-1-GCSGD_2020).S10.S22

Regalado, A. (2015). DARPA Program Seeks to Use Brain Implants to Control Mental Illness. MIT Technology Review. http://www.technologyreview.com/ news/527561/military-funds-brain-computer-interfaces-to-control-feelings/

Rêgo, G. G., Gonçalves, Ó. F., & Boggio, P. S. (2022). Attention neuroenhancement through tDCS or neurofeedback: A randomized, single-blind, controlled trial. *Science and Reports, 12*, 17613. https://doi.org/10.1038/s41598-022-22245-6

Rommelfanger, K. S., Pustilnik, A., & Salles, A. (2022). Mind the gap: Lessons learned from neurorights. *Science & Diplomacy*. https://doi.org/10.1126/scidip.ade6797
Salles, A., & Wajnerman Paz, A. (2024). Anthropomorphism in social AIs: Some challenges. In: Ienca, M., & Starke, G., (Eds.), *Brains and machines: Towards a unified ethics of AI and neuroscience*, (pp.101–118). https://doi.org/10.1016/bs.dnb.2024.02.007
Salles, A. (2018). Neuroethics in context: The development of the discipline in Argentina. In Johnson, L. S. M., & Rommelfanger, K. S., (Eds.), *The Routledge handbook of neuroethics* (pp. 429–441). Routledge/Taylor & Francis Group. https://doi.org/10.4324/9781315708652-32
Salles, A. (2023). Some reflections on the neurorights debate. In *The risks and challenges of neurotechnologies for human rights* (pp. 35–38). UNESCO.
Sample, M., Aunos, M., Blain-Moraes, S., Bublitz, C., Chandler, J. A., Falk, T. H., & McFarland, D. (2019). Brain-computer interfaces and personhood: Interdisciplinary deliberations on neural technology. *Journal of Neural Engineering, 16*, Article 063001.
Sententia, W. (2004). Neuroethical considerations: cognitive liberty and converging technologies for improving human cognition. *Annals of the New York Academy of Sciences, 1013*, 221–228. https://doi.org/10.1196/annals.1305.014
UN Human Rights Council. (2012). *The promotion, protection and enjoyment of human rights on the Internet : resolution/adopted by the Human Rights Council.* (2012, July 16). United Nations Digital Library System. https://digitallibrary.un.org/record/731540?ln=en&v=pdf
UN Human Rights Council. (2014). *The promotion, protection and enjoyment of human rights on the Internet : resolution/adopted by the Human Rights Council.* (2014, July 14). United Nations Digital Library System. https://digitallibrary.un.org/record/775322?ln=en&v=pdf
UN Human Rights Council. (2016). *The promotion, protection and enjoyment of human rights on the Internet : resolution/adopted by the Human Rights Council.* (2016, July 1). United Nations Digital Library System. https://digitallibrary.un.org/record/845727?ln=en&v=pdf
UN Human Rights Council. (2018). *The promotion, protection and enjoyment of human rights on the Internet : resolution/adopted by the Human Rights Council.* (2018, July 5). United Nations Digital Library System. https://digitallibrary.un.org/record/1639840?ln=en&v=pdf
UN Human Rights Council. (2021). *The promotion, protection and enjoyment of human rights on the Internet : resolution/adopted by the Human Rights Council.* (2021, July 13). United Nations Digital Library System. https://digitallibrary.un.org/record/3937534?ln=en&v=pdf
UNESCO. (2023a). *The risks and challenges of neurotechnologies for human rights.* Available at https://unesdoc.unesco.org/ark:/48223/pf0000384185
UNESCO. (2023b). *Neurotecnologías y derechos humanos en América Latina y el Caribe: desafíos y propuestas de política pública.* Available at https://unesdoc.unesco.org/ark:/48223/pf0000387079
United Nations. (2021). *Our Common Agenda: Report of the Secretary-General.* United Nations, New York. Available at https://www.un.org/en/content/common-agenda-report/assets/pdf/Common_Agenda_Report_English.pdf
Véliz, C. (2021). *Privacy is power. why and how you should take back control of your data.* Bantam Press.
Verbaarschot, C., Tump, D., Lutu, A., Borhanazad, M., Thielen, J., Van Den Broek, P., Farquhar, J., Weikamp, J. G., Raaphorst, J., Groothuis, J. T., & Desain, P. (2021). A visual brain-computer interface as communication aid for patients with amyotrophic lateral sclerosis. *Clinical Neurophysiology, 132*, 2404–2415. https://doi.org/10.1016/j.clinph.2021.07.012
Wander, J. D., & Rao, R. P. (2014). Brain-computer interfaces: A powerful tool for scientific inquiry. *Current Opinion in Neurobiology, 25*, 70–75. https://doi.org/10.1016/j.conb.2013.11.013
Yuste, R., Goering, S., Arcas, B. A. Y., Bi, G., Carmena, J. M., Carter, A., & Wolpaw, J. (2017). Four ethical priorities for neurotechnologies and AI. *Nature, 551*, 159–163. https://doi.org/10.1038/551159a
Yuste, R., Genser, J., & Herrmann, S. (2021). It's time for neuro-rights: new human rights for the age of neurotechnology. *Horizons: Journal of International Relations and Sustainable Development, 18*, 154–165. https://www.jstor.org/stable/48614119

Zehr, E. P. (2015). The potential transformation of our species by neural enhancement. *Journal of Motor Behavior, 2015*(47), 73–78.

Zhang, X., Ma, Z., Zheng, H., Li, T., Chen, K., Wang, X., Liu, C., Xu, L., Wu, X., Lin, D., & Lin, H. (2020). The combination of brain-computer interfaces and artificial intelligence: applications and challenges. *Annals of Translational Medicine, 8*, 712. https://doi.org/10.21037/atm.2019.11.109

Advances Towards a Legal Theory of Neurolaw Based on the Case of Mexico

Enrique Cáceres Nieto and Carmen Patricia López Olvera

Abstract This chapter presents the latest on neurorights in Mexico. The first section analyses the conceptual scaffolding of "neurolaw" based on its different meanings in ethics and law, as well as its bearing on "neurorights."The second part explores three main legal positions on the regulation of neurorights: (1) an exclusive one, which does not call for specific categories since they are protected by existing positive law, especially at the international level; (2) an inclusive one, which sees the need for new regulatory frameworks given their specific characteristics and their relation with neurotechnologies and artificial intelligence; and (3) a hermeneutic position, which wants to adapt existing law to these new conditions through interpretation. A brief overview is also given of conceptual issues surrounding the three neurorights at the centre of current discussion -mental integrity, mental privacy and cognitive liberty— as seen from each position. The third section reports on neurolaw-related activities in Mexico to contextualize those specifically associated with neurorights and looks at related constitutional reforms currently under discussion. Lastly, a general theory of law is proposed to help guide academic discourse, legislation and case law work on neurorights both nationally and internationally.

Keywords Experimental jurisprudence · Naturalised legal theory · Neurolaw · Legal epistemology · Contemporary legal philosophy

This paper is the result of research conducted at the Laboratory of Legal Constructivism, Cognition, Complexity and Law, directed by Enrique Cáceres Nieto at the National Autonomous University of Mexico Legal Research Institute.

E. C. Nieto (✉) · C. P. L. Olvera
National Autonomous University of México Legal Research Institute, Maestro Mario de La Cueva S/N Ciudad Universitaria, Ciudad de México, México
e-mail: encacer@unam.mx

P. López-Silva (ed.), *Contextualizing Neuroprotection: Latin American Perspectives on the Impact of Neurotechnological Development in Life and Society*, Ethics of Science and Technology Assessment 54, https://doi.org/10.1007/978-3-031-96055-0_8

1 Introduction

As in all periods of history, science and technology have the power to radically transform the world as one knows it, and today is no exception. Scientific and technological progress over the last years:

> Such as new developments in hemodynamics, electrophysiological imaging and techniques to perform sophisticated, precise analyses on brain lesions; the ability to conduct non-invasive focused brain stimulation, such as transcranial magnetic stimulation; the inclusion of new imaging techniques for neuroendocrinology and genetics like functional magnetic resonance or positron emission tomography; new techniques to view functional and anatomical neuronal connectivity (Cáceres-Nieto & López-Olvera, 2022, 66).

Are leading us to rethink what human beings are.

The impact of neuroscience has broken through the confines of hard sciences and flowed into various social fields, giving way to new interdisciplinary domains like neuro-economics, neuro-marketing, neuro-finances, neuro-pedagogy, neuro-politics, neuro-ethics and, as seen in this book and related research, neurolaw. The three main aims of this chapter are: (1) to contribute to clearing up some misperceptions between neuro-ethics, neurolaw, and neurorights; (2) to describe the current academic and legislative work on neurolaw and neurorights in Mexico, based on the above concepts; and (3) to provide a theoretical and methodological framework to guide legislative work on neurolaw and neurorights at both constitutional and infra-constitutional levels.

2 From Haze to Clarity

Paraphrasing Henri Atlan's metaphor in his "*Entre le cristal et la fumée. Essai sur l'organisation du vivant*" (Atlan, 1979), we hold that knowledge does not follow a sequence of steps but passes from one step to another going from a haze to clarity. In other words, it gradually materializes from a series of vague and unconnected assumptions that build themselves up little by little until they reach a clear and solid state. This metaphor is particularly helpful in the preliminary stages of developing knowledge, as in the case of neurolaw and its relationship with neurorights.

The path to clarity is often fraught with difficulties, especially if we consider that different languages can entail plays on words that lead to certain difficulties, even if referring to the same phenomenon or aspect. This happens, for instance, when the Spanish-speaking world tries to explain certain terms to the English-speaking world.Despite the linguistic differences, we believe there are some common problems and clarifying them might contribute to the advancement of a conceptual ontology that defines this interdisciplinary field known as neurolaw in English and *neuroderecho* in Spanish.

One of the main points of misunderstanding is found in the concepts of neuro-ethics, neurolaw and neurorights, which stem from other more fundamental confusions between law and morality, central themes of traditional legal philosophy. Far from being just an academic problem, confusion between the concepts of law and morality can have serious practical consequences due to moral relativism, which implies that a person may not feel bound to positive law because its content is not compatible with their moral intuition. On the other hand, this same misunderstanding is counterintuitive because in social practices and regardless of moral considerations, people quote the law in the courts and not the Ten Commandments even while both prohibit murder.

In differentiating law and morality, one vital point lies in the constitutive properties of legal provisions; that is, the possibility to determine the conditions for certain legal or institutional facts to exist. This attribute is compatible with the difference between constitutive rules, regulative rules, brute facts and institutional facts proposed by philosopher John Searle (Searle, 1997) who holds that certain facts can only exist in the world if their existence has been previously established by certain linguistic representations that are, to be precise, constitutive laws. One prime example of this is the game of chess: we can identify chess-like facts in the real world because the game has already been built around certain constitutive rules. Something similar happens in law, which can be seen as a system of constitutive rules of institutional facts. Hence. In the world of law, we can take the same type of brute facts and make distinctions between intentional homicide, manslaughter, premeditated murder, and so on.

Unlike law which requires theories from institutional sources to create the rules of the system, rules of morality emerge from social interactions that are not based on constitutive rules. From a systemic perspective, everything that happens within the dimension of law-related institutional facts can be considered intra-systemic because it takes place within a framework provided by the constitutive rules of law. Meanwhile, everything that comes before a law was created can be considered extra-systemic.

The relationship between law and morality is made clear by understanding the concepts framing constitutive, extra-systemic and intra-systemic aspects. Since many moral decisions take place in the extra-systemic area of law, they can become part of the law. When they are formalised through the constitutive rules of the system, they go through the legal extra-systemic scope of the law and, therefore, new law-related institutional facts can no longer exist. For instance, femicides did not legally exist until recently; there were only homicides, regardless of whether the act was committed against men or women. It was only after this new type of crime was instituted that it was possible to find criminal acts consistent with femicide, as well as the corresponding punishment for such behaviour.

The above considerations are important because the lack of a clear differentiation between the normative world of morality and of the law affects the lack of clarity surrounding the operation of both normative systems and the relationships between them. One immediate effect is that papers on neuro-ethics are frequently confused with papers on neurolaw and neurorights without clearly marking the difference as to whether they refer to the province of one or the other.

In view of this, papers today that contain the word "neurolaw" in their title often fall under the province of ethics and not law. Thus, the lack of conceptual clarity about law and morality can lead to confusion between studies on neurolaw in the ethical sense and neurolaw in the legal sense. It further complicates things when attempting to differentiate neurorights from a moral perspective and neurorights from a legal perspective. The fact that these confusions still surface at the current stage of neurolaw and neurorights research is apparent in some of the definitions proposed for "neurolaw" in the academic sphere:

[...] the analysis of the law and justice from the perspective of neuroscience, which implies an understanding of human behaviour through the study of the brain and its interaction with its environment. (García-López, 2021).

[...] the study of the ethical, legal and social issues that arise when scientific findings on the brain shift to medical practice, legal interpretations and health and social policies (Marcus, 2002).

It must be stressed that within the conceptual ontology of neurolaw, expressions like "neurorights" in English or "*neuroderechos*" in Spanish proposed by Dr Rafael Yuste and the Morningside group as a new category of human rights (Yuste et al., 2017) fall under the extra-systemic nature of positive law and are considered what is philosophically known as regulative ideals. However, they are still in the process of moving from the extra-systemic to the intra-systemic region of law in view of reforms, constitutional or otherwise, that have been enacted in some parts of the world, as will be seen below.

After unravelling the relationship between neurolaw and neurorights in the fields of ethics and of law, and considering the lack of a legal definition for either one, Cáceres proposes the following definition of neurolaw:

> [...] refers to the interdisciplinary field that studies the neurocognitive processes of legal operators (neurolaw 1), as well as the role of neurosciences as a way to enhance the law (neurolaw 2), the legal oversight of neuroscientific research (neurolaw 3), and the protection of the subjective rights associated with the neurological activity of individuals (neurorights 4) (Cáceres-Nieto et al., 2021, 79).

The first entry could also be called the neuroscience of legal operators since 'neurolaw' is considered a branch of social neuroscience that plays a role like that of neuro-ethics or neuro-economics.

The second definition corresponds to forensic neuroscience, which is intricately linked to applied legal epistemology and the problems surrounding the determination of the truth in the law, as might be the case when it is necessary to determine whether a person can be imputed or acted in bad faith.

The third entry refers to what could be called the law for neurosciences aimed at regulating neuroscientific research.

The fourth definition essentially deals with the ethics of neurosciences alluding to the law protecting neurorights, i.e., as an extension of positive human rights (Cáceres-Nieto et al., 2021, 79).

3 The Descriptive Model of Neurolaw-Related Issues Being Discussed in the Mexican Epistemic Community

3.1 Methodology

The methodology used for the descriptive model of neurolaw in Mexico and in general can be summarized as follows:

An analysis was made of the written works (books, chapters and articles) and several recorded academic events on neurolaw in general to identify the key descriptors of the issues addressed in each, yielding a total of 61 (Cáceres-Nieto & López-Olvera, 2024). Close attention was given not only to the titles, but also to the content itself since titles do not always reflect the contents of the study and sometimes even examine important concepts that are not mentioned in the title. The list gathered includes issues studied by published Mexican and foreign authors and presented at events organized in Mexico.

Google Scholar and BidiUNAM were used to locate written papers while the Google search engine was used to search for academic events in general (seminars, conferences, congresses, forums, etc.) and on social media platforms, specifically Facebook and X. The identified sources and descriptors were saved on an Excel spreadsheet (Cáceres-Nieto & López-Olvera, 2024).

The following phase consisted of creating general categories for each of the four entries of "neurolaw" posited by Cáceres: behavioural endocrinology, teaching neurolaw, neuroscience applied to understanding the neurocognitive processes of legal operators, neuro-criminology, neurorights in general, neurorights from a legal perspective, specific neurorights, and legal neuro-epistemology (Cáceres-Nieto & López-Olvera, 2024).

3.2 Findings

The frequency distribution of the findings is shown in subsets on the following graph (Fig. 1):

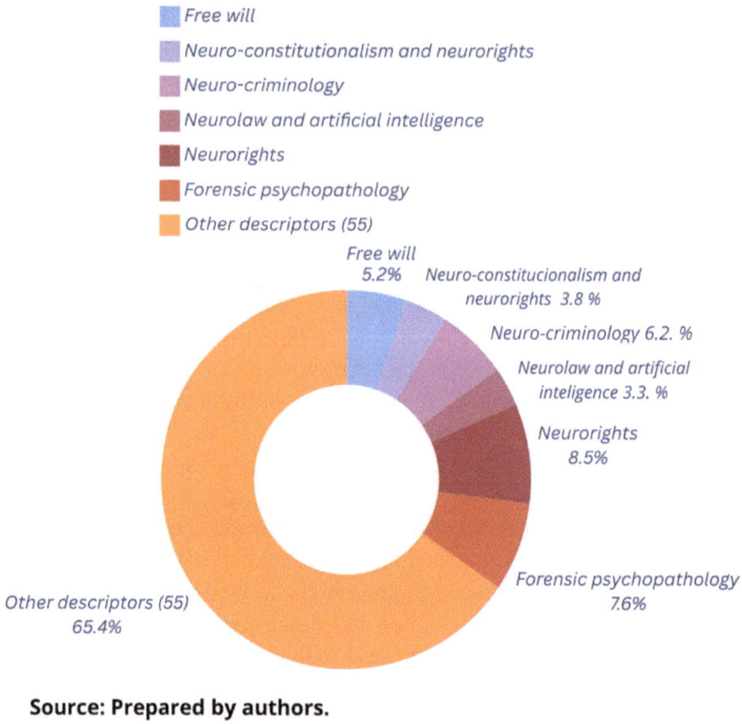

Fig. 1 Most frequent descriptors in publications and academic events with 'Neurolaw' in Mexico. *Source* Prepared by authors

Graph 1 shows that neurorights in general has drawn the most attention in the epistemic community on neurolaw, followed by forensic psychopathology, neuro-criminology, free will, neuro-constitutionalism and neurolaw, and neurolaw and artificial intelligence.

The set of descriptors in the more general categories that correspond to the entries of "neurolaw" included in Cáceres's definition are shown in the following graph (Fig. 2):

Graph 2 reveals that most works fall under the category of neurolaw in general and are consequently found within the realm of the extra-systemic area of law. Moreover, most of these deal with observations on specific neurorights.

As to "neurolaw" in its law-related entry, most research focuses on "neurosciences as a way to enhance the law" as part of neuro-criminology (which includes forensic neuropsychiatry, forensic neuropsychology, etc.). Research in other areas in this category of neurolaw are negligible despite its importance and is limited to categories of neurolaw that centre on its legislation, legal neuro-epistemology and neuroscience applied to understanding the neurocognitive processes of legal operators. These findings shed light on the areas that should be studied in more depth.

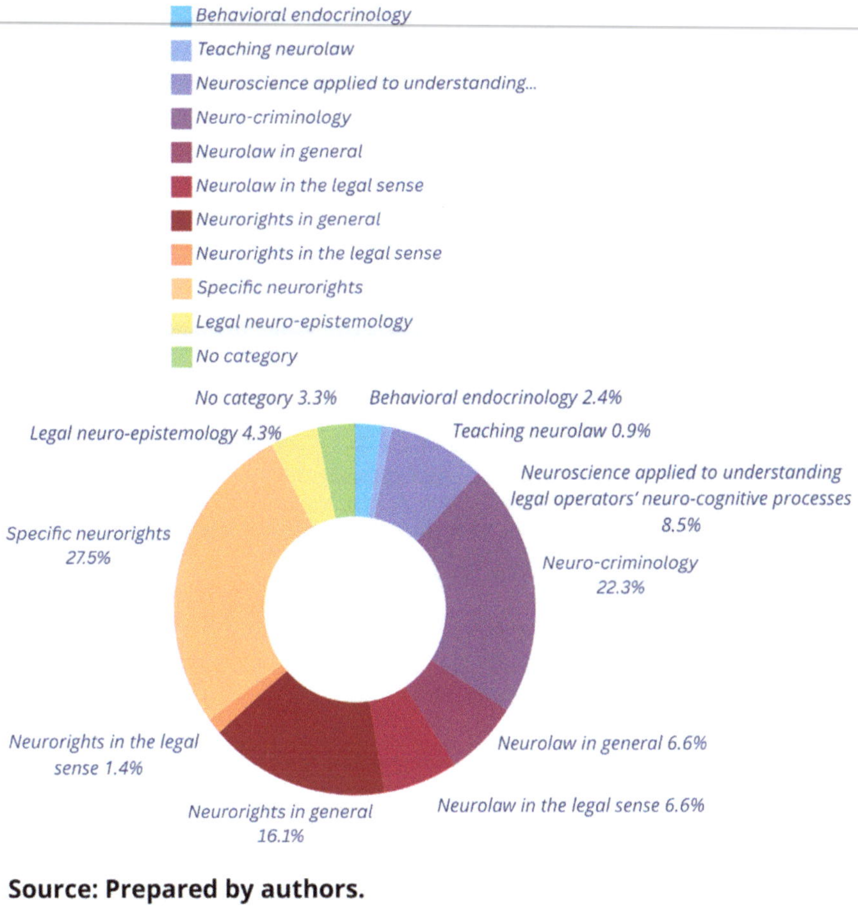

Fig. 2 Categories and percentages in publications and academic events associated with 'Neurolaw' in Mexico. *Source* Prepared by authors

4 Legislative Activity Concerning Neurorights in Mexico

Like in most countries, legislative activity on neurorights is in its initial stages. Even then, two initiatives for constitutional reforms have been introduced so far: one by the Chamber of Deputies and the other by the Senate.

On 1 August 2023, the Chamber of Deputies proposed the following changes to the Mexican constitution:

Article 4.—Man and woman are equal under the law. The law shall protect the organization and development of the family. […]

Text to be added: Every person has the right to a full and integral individual identity, as well as physical and psychological integrity as a condition for freedom. The State shall guarantee respect for the privacy and mental integrity of persons.

No authority or individual may modify, reduce or affect this integrity or identity by means of any technological device. (Draft Decree to Add a Ninth Paragraph to and Recognize the Subsequent Paragraphs of Article Four of the Political Constitution of the United Mexican States, 2023).

Meanwhile, the initiative presented by the Senate proposes an amendment to a different constitutional article referring to the powers of Congress, which currently reads:

Section 17. Enact laws on general communication media, information and communication technologies, radio broadcasting, telecommunications, including bandwidth and internet, posts and emails, and on the use and exploitation of waters under federal jurisdiction. (Article 73, Section XVII, of the Political Constitution of the United Mexican States).

The Senate has proposed that it read:

> Enact laws on general communication media, information and communication technology, radio broadcasting, telecommunications, including bandwidth and internet, as well as on artificial intelligence and its applications, cybersecurity, neurorights, posts and emails, and on the use and exploitation of waters under federal jurisdiction (Draft Decree to Reform Article 73, Section XVII, of the Political Constitution of the United Mexican States, on Matters of Artificial Intelligence, Cybersecurity and Neurorights 2023).

The basic difference between these bills is that while the one from Chamber of Deputies is about a reform to the content, the Senate's is simply about jurisdiction. Also of note is that the Chamber of Deputies' proposal does not explicitly include the protection of the neurorights proposed by Dr Rafael Yuste and his research team.

At the time of writing this chapter, the future of both projects is in the air, especially in view of the change of government administrations and the newly elected members of Congress. Work on this is expected to resume in the near future with collaboration of UNESCO Mexico, which has recently begun work on creating ethical guidelines on neurolaw and neurotechnologies.

5 Neurorights from a Legal Perspective as Conditions of Normative Effectiveness and Second Level Subjective Rights

To summarize, an analytical study was carried out to establish where exactly these concepts are positioned in the domains of morality and law. A definition for "neurolaw" from a legal perspective, which includes neurorights, was given. An empirical model of the status of neurolaw and neurorights in Mexico was provided to identify the issues that have garnered more attention, highlighting the fact that neurorights is no longer the focal point of most research being carried out. Reference was also made to national legislative work on neurorights and its relationship with artificial intelligence.

In keeping with position of neurorights in extra-systematic and intra-systematic areas, the following is proposed to contribute to incorporate these rights and institutional protections intra-systematically:

The legal status of neurorights presents four basic positions on their regulation: (1) an exclusive one that does not consider it necessary to create specific categories for neurorights since they are already protected by existing positive law, particularly at the international level; (2) an inclusive one that sees the need for establishing new categories and a regulatory framework for neurorights, given their specific characteristics and their ties to neurotechnologies and artificial intelligence; (3) a hermeneutic one that would adapt existing laws to the new conditions by means of normative interpretation; and (4) a progressive one that, while at a level of normative abstraction neurorights are protected-to a certain degree-under the umbrella of human rights, their specific features and dynamics need to allow for future legislative and case law-related developments when artificial intelligence and neurotechnologies are involved. This chapter stands by this last position.

As mentioned above, most deliberations on neurorights in moral or political philosophy circles place them in the extra-systematic area of the law. To turn these neurorights into constitutive rules of a legal system, it is necessary to reconsider their logical and normative structure in such a way that it fulfils what Cáceres calls "conditions of normative effectiveness", i.e., to incorporate them as cogs in the machinery that guarantees effective processes pertaining to law-related institutional facts (Cáceres-Nieto, 2024) .

From an analytical perspective, the normative system is made up of definitions and prescriptive statements qualified in deontic terms as prohibitive, mandatory or granting subjective rights. These last can be divided into two groups: rules that establish subjective rights of action and rules that establish subjective rights to enjoy states of affairs. The rules for the first type of subjective rights include, for example, the right to buy, vote, get married and so on. Some examples of rules for the second type of rights are the right to a healthy environment, the right to health, the right to peace and so on. The deontic structure of these rules, founded on the bases given by the deontic operator and behaviour, always refer to a normative subject.

The concept of conditions of normative effectiveness refers to the fact the law needs to be linked with various legal provisions to ensure its effective operation. Hence, for Subject "S1" to be entitled to a subjective right contingent on the occurrence of a given condition, it must be tied in with the existence of another rule that obligates Subject "S2" to perform the act upon which the fulfilment of said condition depends. In a technical sense, this obligation must have another rule setting a penalty for S2 if they do not fulfil their duty toward S1. For S2 to be penalized and, if applicable, to fulfil their obligation toward S1, a subjective right must be available so that S1 can file legal action before the corresponding institution if S2 does not comply.

The above framework for neurorights show that they are Type 2 subjective rights since their conditions correspond to mental privacy, identity and personal autonomy, free will and self-determination, equal access to a cognitive approach, and protection against algorithmic biases, to name a few.

This means that shifting these rights to the intra-systemic area of law, especially to the level of infra-constitutional legislation, requires other normative cogs to ensure they will be effectively upheld.

We propose some guidelines for positing neurorights as second-generation rights as they correspond to rights that depend on specific conditions. It should be noted that in an effort to protect brain and neurological life to the highest possible extent from the improper use of neurotechnologies and artificial intelligence, as well as to dispel any doubts regarding the ownership of neurorights, they should be explicitly regarded as part of the rightsholders' right to ownership.

5.1 Level 2 Neurorights

All data obtained through neurological and/or endocrine activities, especially those occurring in the brain, are the natural property of the humans in whom they occur.

Closely behind the right to natural ownership of neurological or endocrine activity, as well as of the data and information that can be obtained from them, are other rights to ownership, such as (1) the right to grant access to one's physiological activity; (2) the right to allow the data collected from said activity to be recorded; (3) the right to allow these data to be stored; (4) the right to allow the data to be transformed into information; (5) the right to authorize said data and information to be processed by algorithms; (6) the right to authorize the type of outcomes from neurotechnologies and artificial intelligence; (7) the right to authorize the end use to be given to these data, especially when it might affect first-generation neurorights either directly or potentially; and 8) the right to withdraw consent to third-party use of any or all of the above neurorights.

5.2 Mandatory Rules to Protect First and Second-Generation Neurorights

No individual or corporation can have access to record, process, transform, interpret or use neuro-data without the consent of their natural owner.

The deontic status of second-generation neurorights is that of subjective rights for its natural owner.

The deontic status of third-party behaviours regarding access, registry, transformation, algorithm processing and use of the results of this processing implies the obligation of respecting and the prohibition of disregarding either generation of neurorights.

For it to be legally valid, the owners' exercise of the above neurorights entails an explicit contractual relationship (sale, donation, rental, etc.) when third parties are involved and is to be governed by freedom of choice.

The exercise of these rights in favour of third parties modifies the deontic status of prohibited behaviours that give way to individualized rules of permitted ones, i.e., as long as the natural owner of the second-generation neurorights does not withdraw their consent to third-party use of their subjective rights.

In the case of algorithm processing, especially when done with artificial intelligence, third parties will be subject to the existing legal provisions for developing systems of this type. If there is a legislative gap, should be applied the European Union Regulations and the UNESCO guidelines for human rights protection, especially when it concerns high-risk artificial intelligence systems, with normative soft law status. Thus, the principle to be applied is the one that ensures the fulfilment of the secondary obligations required to comply with the main obligation.

Third-party exercise of the rights obtained from the subjective rights of their natural owners does not exempt third parties from legal (and particularly criminal) responsibility.

Considering the distinct features of the different types of neurotechnologies, it is necessary to stipulate the types of behaviours that constitute good practices and those that should be prevented by means of official rules.

Each State should flesh out the rights and obligations vaguely outlined here depending on their public policies, institutional designs and regulatory architecture, aiming at all times to ensure the highest level of protection for the first-level neurorights proposed by Dr Yuste.

References

Atlan, H. (1979). Entre le cristal et la fumée. Essai sur l'organisation du vivant. Seuil.

Cáceres-Nieto, E. (2024). La Teoría Sintáctico-Constructivista: Un Ensayo sobre la Renovación de la Dogmática Jurídica. In F. Castaños, J. Nadal Palazón, & M. Palacios (Eds.), Resignificaciones: lenguajes en acción (1st ed., Vol. 1, pp. 443–477). Instituto de Investigaciones Filológicas.

Cáceres-Nieto, E., Diez García, J., & García García, E. (2021). Neuroética y neuroderecho. Revista Del Posgrado En Derecho de La UNAM, 15, 37–86. https://doi.org/10.22201/ppd.26831783e.2021.15.179

Cáceres-Nieto, E., & López-Olvera, C. P. (2022). El neuroderecho como un nuevo ámbito de protección de los derechos humanos. Cuestiones Constitucionales. Revista Mexicana de Derecho Constitucional, 46, 65–92. https://doi.org/10.22201/iij.24484881e.2022.46.17048

Cáceres-Nieto, E., & López-Olvera, C. P. (2024). Literature Review on NeuroLaw and Neurorights in Mexico. https://docs.google.com/spreadsheets/d/1MPsVVARPa29H2yuRNqk0Iyp6Qcb4pUxY/edit?usp=drive_link&ouid=101620914262486682385&rtpof=true&sd=true

Draft Decree to Add a Ninth Paragraph to and Recognize the Subsequent Paragraphs of Article Four of the Political Constitution of the United Mexican States, 2023.

Draft Decree to Reform Article 73, Section XVII, of the Political Constitution of the United Mexican States, on Matters of Artificial Intelligence, Cybersecurity and Neurorights, 2023.

García-López, E. (2021). Derecho penal y neurociencia (E. García-Lopez, Ed.; Edicion Ki). Instituto Nacional de Ciencias Penales.

Marcus, S. J. (2002). Neuroethics: Mapping the field. Dana Press.

Searle, J. R. (1997). *The construction of social reality*. Free Press.

Yuste, R., Goering, S., Arcas, B. A. Y., Bi, G., Carmena, J. M., Carter, A., Fins, J. J., Friesen, P., Gallant, J., Huggins, J. E., Illes, J., Kellmeyer, P., Klein, E., Marblestone, A., Mitchell, C., Parens, E., Pham, M., Rubel, A., Sadato, N., & Wolpaw, J. (2017). Four ethical priorities for neurotechnologies and AI. *Nature, 551*(7679), 159–163. https://doi.org/10.1038/551159a

The Right to Health in Brazil and the Prospects for Using Neurotechnology

Gabrielle Bezerra Sales Sarlet and Lucas Reckziegel Weschenfelder

Abstract This article explores the evolving legal framework governing neurotechnology within Brazil's digitalized health system. The analysis begins with an examination of Brazil's healthcare digitalization policies, emphasizing the critical role of data protection as a foundational and reflective framework in this context. It then provides an overview of state policies and regulatory frameworks related to telemedicine and human research, highlighting their potential adaptations to accommodate neurotechnologies. Finally, the article addresses the emerging perspective on neurorights, with a focus on Brazilian and broader Latin American legal initiatives. The discussion concludes by underscoring the imperative of a cross-updating of the regulatory system, particularly concerning healthcare, medical practice, and research, with specific attention to neurotechnologies incorporating artificial intelligence.

Keywords Right to health · Neurotechnology · Neurorights · Brazil Constitutional Law

1 Introduction

In contemporary times, society is undergoing a massive process of computerization, digitalization, and algorithmization. The concept of ubiquitous computing, envisioned by Weiser (1993, 72) during the 1980s and 1990s, has, in a certain sense, materialized. Modern society finds itself in a transitional era, shifting from traditionally physical and analog structures to a more cyber-physical and digital framework.

The health subsystem is no exception to this transformation. This text aims to examine these changes, primarily through a legal lens, focusing on Brazil's constitutional health subsystem, with an emphasis on the human and fundamental right to

G. B. S. Sarlet (✉) · L. R. Weschenfelder
Pontifícia Universidade Católica do Rio Grande do Sul, Porto Alegre, Brasil
e-mail: gabrielle.sarlet@pucrs.br

L. R. Weschenfelder
e-mail: lucas.reckziegel@edu.pucrs.br

P. López-Silva (ed.), *Contextualizing Neuroprotection: Latin American Perspectives on the Impact of Neurotechnological Development in Life and Society*, Ethics of Science and Technology Assessment 54, https://doi.org/10.1007/978-3-031-96055-0_9

health (as it is operationalized in Brazil), and the impact of emerging technologies, particularly neurotechnologies. These technologies are viewed within the framework of constitutional law, specifically in relation to the state's duty of protection, as defined by the doctrines of the prohibition of insufficient protection and the prohibition of excess, originally from German legal thought and applicable in Brazil.

This paper seeks to present, in three sections, reflections and mappings of the initiatives that are currently being implemented or projected in Brazil within the neurotechnological landscape. These initiatives are addressed not only from a regulatory perspective in the healthcare sector but also in light of the emerging grammar of so-called neuro-rights. This is particularly important given that the recent 2023 ICT Health (2023)[1] survey indicated a relatively slow adoption of artificial intelligence systems (AI systems) in the health sector, despite significant progress in online services, telehealth, treatments, and diagnostic assistance utilizing neurotechnologies, with or without embedded AI.

The first part provides a concise overview of the digitalization process in Brazil's health sector and the associated challenges to the right to health, recognized as a fundamental right. It also examines how the protection of personal data, enshrined as a fundamental right, serves as a pivotal axis of analysis in this context—particularly at the intersection of health and neurotechnologies. The second section offers observations on the intersection between neurotechnological advancements and the legal-regulatory framework for healthcare in Brazil, in the structures of telemedicine and research with human beings, for the development of technological instruments. Mention is made of both regulatory orders, considering their normative nuances that acclimatise perspectives for the adoption and development of neurotechnologies in this specific scenario, which is still under construction, highlighting both the potential benefits and the associated risks and challenges. The objective of this section is to elucidate the existing legal framework on the subject, contextualized within the backdrop of health digitalization in Brazil. It is emphasized that, through an interpretative approach, these legal provisions may be applied to the field of neurotechnology, particularly in the context of medical practices and research development. However, this section, in connection with the third and final part, highlights the need for a more comprehensive analysis of this emerging technology, focusing on its distinct areas of application and their respective legal requirements, including those related to research, medical practice, and consumer use, among others. Finally, the third section situates this neurotechnological scenario within Brazil's constitutional and

[1] The survey found interesting reasons for not using artificial intelligence in healthcare. Use was not a priority (63 per cent), incompatibility with existing equipment, software and hardware (52 per cent), costs would be too high (50 per cent), lack of need (49 per cent), difficulty finding information on the quality or quality of the data needed to use the technology (47 per cent), lack of training for professionals to use the technologies (46 per cent), concerns about privacy and data protection (39 per cent) and ethical problems (22 per cent). In the case of the use of AI systems themselves, the following were identified: to improve digital security (45%), assistance in organisational processes (41%), improve the efficiency of treatments (38%), support in logistical processes (37%), assistance in diagnostics (33%), support in human resources (28%).

legal landscape, identifying specific concerns raised by current initiatives. Additionally, there is a brief comparison with similar proposals in Latin America, stressing the importance of considering these initiatives within the broad context of fair, secure, robust, and reliable AI modules when embedded in neurotechnologies.

2 The Digitalization of Healthcare in Brazil and the Importance of Personal Data Protection Regulation

This section aims to analyze the extensive state reform processes underway in the Brazilian constitutional framework, which intersect with the digitalization of health and, consequently, with the ways in which the right to health—recognized as a fundamental social right in Brazil—must adapt. The discussion underscores the significance of the doctrine of personal data protection in this context, framing it as an emerging form of self-reflexivity. Furthermore, the section seeks to highlight the evolving landscape of digitalization within which neurotechnologies—primarily in the realms of medical practice and research—are already being utilized and conceptualized for functional and future applications. In essence, this represents a trajectory that is actively taking shape.

In Brazil's constitutional framework, re-established through the democratic process in 1988, a transformation of both society and the state was initiated with the promulgation of the Constitution of the Federative Republic of Brazil (1988) (hereinafter CF/88). A federative structure was established, within which the responsibilities and competencies of the Union, Federated States, and Municipalities (and Federal Territories) are delineated, sharing the duty to guarantee the right to health, "through social and economic policies aimed at reducing the risk of disease and other health hazards, and ensuring universal and equal access to actions and services for its promotion, protection, and recovery" (Article 196, CF/88). This structure is closely tied to the broader constitutional framework of social fundamental rights, which also includes the statement that "social rights include education, health [...]" (Article 6, CF/88).

To illustrate this structure, mention can be made of Law No. 8.080/1990, which regulates healthcare actions and services across the national territory, whether carried out individually or jointly, on a permanent or temporary basis, by individuals or legal entities under public or private law (Article 1). The right to health in Brazil is founded on the structuring principle of universality, ensuring access to healthcare for all Brazilian citizens and migrants, irrespective of any prior contributions. Moreover, it is recognized as a fundamental right that may be enforced through judicial mechanisms (doctrine of prestational rights), subject to specific criteria and contingent upon the particular domain of healthcare involved. Nonetheless, this universal right coexists with the disparities that characterize the dual structure of the public and private healthcare sectors, with the latter operating on a contributory basis.

This right, with its fundamental and multidimensional nature, represents a diffuse foundation that shapes both the state and society, impacting over 200 million people across the country's vast territory. Along with the constitutional structure that underpins it, has not remained unaffected by the current digitalization of life, a phenomenon already foreseen in similar terms by the concepts of ubiquity (Weiser, 1993)[2] and computational power. It is important to note that in Brazil, there is a constitutional provision regarding science, technology, and innovation, particularly in Articles 218 to 219-B, with Article 218, § 1°, emphasizing that basic and technological scientific research shall receive priority treatment from the state, with a view to public welfare and the advancement of science, technology, and innovation.

A range of state and private reforms are currently underway, quietly revolutionizing both the digital[3] (Karaboga et al., 2016, 32) and material aspects of traditional forms of life, including areas where the state is (or should be) more concretely involved. This process, however, is not without criticism, particularly regarding the existence of certain pervasive and interventionist strategies[4] (Steinmüller, 1975, 514–515), as can be seen in the legal and political handling of healthcare and the parallel process of datafication. This is not to criticize the real possibilities that datafication brings to administrative organization and management,[5] as well as research and healthcare itself, through computational technologies that digitize traditionally

[2] "The most profound revolutions are not the ones trumpeted by pundits, but those that sneak in when we are not looking. Time will tell if ubiquitous computing, or something else, will be the next to quietly transform our lives" (72).

[3] Of particular note is the profuse data infrastructure that has been established, without presenting it in the public sphere, without getting close to the real advances and objectives, the interoperability of systems and databases, and the risks that such reforms can reveal. These reforms are presented as natural phenomena, in the process of brief state and procedural reforms of transition from analogue to digital. This is even present in recent legislation and policies, which have been advancing more sharply since 2022. Just to give an example, the risks of sensitive personal data being leaked in the field of health of millions, which will occur in 2022.

[4] Deliberately or not, at least in Brazil, there seems to be a practice of reforming and making invisible the computer systems that actually make the administrative and private machinery 'work'. This is a traditional theme in various fields of knowledge, particularly in law, in the dogmatic area of personal data protection. One example is Steinmüller's comment on the automated systems of the 1970s and 1980s in Germany: "Einem wieteren nervensystem vergleichbar, stehen sie "hinter" den traditionellen Verwaltungen, und sind darum für die Bevölkerung relativ unsichtbar, wie Abbildung veranschaulichen soll". Translated by the authors: Similar to another nervous system, they are 'behind' the traditional administrations and are therefore relatively invisible to the population, as illustrated in the figure.

[5] This also produces risks, as has already happened in several countries and in Brazil itself, for example in the automated denial of social security rights in the Netherlands (SyRI case), the discriminatory structure towards women and the poor, as well as in the United States of America in the recent TennCare Connect system case (Deloitte).

In Brazil, due to the ever-increasing number of disputes regarding social rights, the National Council of Justice (CNJ) is in contact with the Social Security Institute in order to promote interoperability aimed at automating the granting of social security rights recognised by the courts.

It is worth highlighting the existence of at least three (registered) cases of leaks of personal data and sensitive personal health data within the digital infrastructure of the Brazilian Ministry of Health. The SCPA case, the CADSUS case, and the E-SUS Notifica and CADSUS cases.

analog practices and create new ones, enhancing progress in the field—especially with neurotechnologies, the central focus of this paper.

By way of illustration, the following can be cited: the Brazilian Strategy for Digital Transformation (E-digital 2022–2026), the Digital Health Strategy for Brazil (2020–2028), the National Digital Government Strategy and the National Digital Government Network (2024–2027), with the priority of implementing digital transformation initiatives for public health policies and services (Article 10, VI), and the National Artificial Intelligence Plan (2024–2028) (which still requires further detailing by the Brazilian state). Among the "immediate impact actions" listed in this plan are those related to healthcare, as well as numerous pieces of legislation that contribute to the advancement of datafication and the digitalization of the constitutional framework under consideration.

It is important to emphasize that these reforms and incentives are and will be operated through the healthcare operational subsystem, namely the Unified Health System (SUS), along with its digital face, E-SUS (the state's organizational structure), and the Meu SUS Digital (formerly Conecte SUS) system. This system not only serves as a structural framework but also as a platform for health in the relationship between the state and the citizen/user, without forgetting the private platforms that co-exist and co-operate with public structures. As a result, it demands a re-reading of the normative provisions of the Charter of Rights of Public Service Users, particularly the Charter of Health Service Users' Rights, for the digital context.

It is no coincidence that in 2022, with Constitutional Amendment No. 115, a reform of the CF/88 text was enacted, adding the fundamental right to the protection of personal data, including in digital environments. This textual change represents a longstanding Brazilian concern, as seen in doctrinal works, scattered legislation, and judicial decisions, which only materialized into a law in 2018, with the Law nº 13.709, of 2018—General Data Protection Law (Lei Geral de Proteção de Dados Pessoais—LGPD 2018) (Sarlet & Ruaro, 2021). This protection has since been enshrined by the decisions of Brazil's Constitutional Court and further strengthened with its incorporation into the constitutional framework.

This constitutional and infraconstitutional process is attentive to the various legal subsystems that are taking shape amid the massive transformations (Albers, 2016) underway. To operationalize their reforms geared towards digitalization, these subsystems will, in one way or another, need to handle and interoperate millions, if not billions, of personal and non-personal data within increasingly centralized information architectures, at least in Brazil, in the public and private sphere.

In the context of technological externalities, particularly digital ones, personal data protection emerges as a legal-normative frontier capable of prompting these processes to reflect on themselves. The dogma of personal data protection has evolved with its historical premises, notably: (i) "[…] information was now devoid of context and thus devoid of control of the subject of the information; (ii) […] availability of data and the technical ability to make use of it created a power imbalance until then unknown […] the core of data protection is to regulate the informational power asymmetry […] and (iii) "[…] it had become clear that the processing of data was not a single act restricted to certain areas of life (Döhmann, 2022, 79)" impacting (or having the potential to impact) individuals in their intersectionalities.

These considerations are compounded by other legal frameworks not covered in this paper but which, in brief, call for attention to other legal modalities associated with the right to health and the right to personal data protection, such as the notion of free development of personality[6] (Ascensão, 2008) (of extreme systemic relevance to Brazil's constitutional subsystem), autonomy and self-determination, and the dignity of the human person (Sarlet, 2015) (Kirste, 2016), which stands as a supreme criterion that structurally intersects with various legal layers involved in the protection of the individual, in an increasingly digitalized society (Hoffmann-Riem, 2021).

This association is significant, not only from a legal-normative reflective standpoint, particularly concerning the growing relationship between law and health but also to ensure that one of the primary functions of personal data protection is upheld: the protection of the subject, the holder of personal data that is extracted, inferred, and reshaped within the asymmetric socio-technological-scientific environment in which they are embedded—prioritizing the individual over the data and information themselves, which are merely their manifestations and by-products. The protection of personal data, in its dogmatic articulation, is part of a normative network concerned with the intersections between the person and the technological structure, notably, in this space, with computer technology. Health data, in the Brazilian legal system, is considered sensitive personal data, demanding greater protective attention, although, as will be realised later, there is a desire to regulate the processing of neural data, as a specific form of data processing, although also, in one of the projects, placing it as sensitive data, connecting dogmatic and regulatory sectors, in their differences, although sharing high levels of protection, depending on the operating context.

The intersection of the accelerated digitalisation of the Brazilian state's information structures—particularly and prominently in the healthcare sector—the relocation of health-related practices and technologies, and the internal reflexivity demanded by the discipline of personal data protection represents an initial step that warrants continued attention and development. This intersection serves as a marker of both subjective and structural dimensions of data within digital environments.

When examined more broadly, especially in relation to multifaceted initiatives involving neurotechnologies in both the public and private spheres, this legal discipline necessitates robust support from the regulatory framework, encompassing both *ex ante* and *ex post* measures. Such regulation must address diverse applications, including health, research, consumerism, wellness technologies, and other domains—many of which occupy materially and legally nebulous spaces.

Furthermore, it is crucial to engage with the meta-theory of human and fundamental rights law, which evolves over time. This dynamic process prompts infra-constitutional legal spheres to reflect on their own practices while considering the broader legal frameworks that underpin them. In Brazil, no comprehensive and specific *ex ante* or *ex post* regulation currently exists in this domain, except for

[6] José Oliveira Ascensão uses the terminology the realisation of personality, widely adopted in Portuguese law.

the regulation of equipment and services for medical purposes, as is the case in the European Union (Medical Devices Directive).

However, there are emerging regulatory microcosms that can—and indeed are—being utilized to operationalize specific medical and healthcare and research practices in the field of neurotechnology. These developments intersect with the broader universal process of digitising healthcare, as outlined in this section.

Considering the implications of these themes, particularly in relation to neurotechnologies, is essential. Thus, the next section will address this topic in greater detail, however, focusing on some regulatory bases that call for reflection and care in the development of this evolving digital sphere.

3 A New Stage: Neurotechnologies and Brazilian Health

The objective of this section is to elucidate the existing legal framework on the subject, contextualized within the backdrop of health digitalization in Brazil. It is emphasized that, through an interpretative approach, these legal provisions may be applied to the field of neurotechnology, particularly in the context of medical practices and research development. The fundamental right to health, in this space, can be interpreted in its multidimensionality, accompanying its holder in preventive practices, medical and assistance relationships, and in research and clinical environments, worlds that are relocated in computerization and digital universes.

What is referred to as neurotechnology represents a distinct form of connection between humans and machines, intensifying the traditional boundaries established by modern scientific and technological evolution. It's important to consider that, although a concept reduces the complexity of the world in order to be operationalised in a given environment, it can also reveal itself as a dogmatic self-delusion of the world and its complexity. For this reason, the concept of neurotechnology adopted in this text, is open and malleable to the complexity of the subject. It refers to devices that, to some extent, incorporate computing in its digital form, in a process of direct/interventionist or indirect/non-interventionist connection with the human brain and nervous system (central and peripheral) (embedded or non-embedded) (wearables, implanted or other forms). One must consider the possibility of complementary or even exclusive and distinct technological forms, such as pharmaceutical services and products, which may be understood as such, as well as technological mechanisms related to cooperation with other parts of the human body, with which, and to the extent that, they indirectly establish links with the brain and nervous system, which are also sometimes called ingestible computers[7] (micro and chemical biocomputers,

[7] In order not to go overboard with explanations, which can be read in specialised literature, an example is cited: Abilify MyCite (aripiprazole tablets with sensor) has an ingestible sensor embedded in the pill that records that the medication was taken. The product is approved for the treatment of schizophrenia, acute treatment of manic and mixed episodes associated with bipolar disorder and for use as an add-on treatment for depression in adults. The system works by sending a message from the pill's sensor to a wearable patch. The patch transmits the information to a mobile

which can connect to the nervous system directly or indirectly, with the possibility of sending messages to an internal or external device).

Regulatory law, as part of the law, is also creative and has the power to create its own ways of operating, regardless of the external environment that affects it. This is a relevant faculty of the subsystem of law, which, in this space, we do not intend to discuss in the epistemological sense, or even in the strict regulatory merits of this performative modality of law. We are therefore aware of the different regulatory spheres that are sometimes created between pharmaceutical services or medicines, for those computer devices, linked to a grammar related to the purposes of the regulated object. We acknowledge that the comprehensive conception of neurotechnology adopted for this text may be subject to criticism and may not be fully embraced within certain regulatory frameworks. It is, therefore, important to clarify that the concept under discussion does not pertain specifically to the internal mechanisms preserved within a localized regulatory legal system for its operationalization. However, this statement emphasizes the need to reflect (which will not be done in this space) on the divisions and boundaries between biology and the computer devices incorporated into the internal workings of regulatory law.

Neurotechnologies enhance the relational possibilities of biotechnologies. They are characterized as "devices and procedures used to access, monitor, investigate, assess, manipulate, and/or emulate the structure and function of the neural systems of natural persons" (OECD, 2019). Furthermore, they may represent extremely problematic forms and designs of informational worlds, beyond the incorporation currently being studied within the Health system (Costa, 2023) and its relationship with the Law system, especially with emphasis on the Brazilian constitutional subsystem.

In the legal dogmatic environment, it has become common to emphasize concerns with these traditional terms: personal data, self-determination, personality development, human dignity, among other frameworks that are effectively open to reproducing themselves to address many social and, in this case, techno-scientific developments, in a constitutional phrasing, particularly in Brazil, stemming from Article

application so that patients can track the ingestion of the medication on their smart phone. Patients can also allow their caregivers and physician to access the information through a web-based portal (FDA, 2017).

5, § 2[8] of the CF/88, which is presented as "a molded form that develops as it lives[9]" (Goethe in the Thought of Peter Häberle, 2024, 11).

With regard to the right to health, this dynamic is also highly relevant. Observing the right to health while assisting in the constitution of the Health system through the incorporation of techno-scientific mechanisms, e.g., neurotechnologies, is part of a process of engaging with normative frameworks that have local, regional, national, and transnational instantiations and intersections.

Within the Brazilian constitutional framework, certain initiatives are emerging that aim to address this scenario of digitalization, though still in an embryonic stage, naturally seeking solutions in an environment characterized more by uncertainty (G'sell, 2024, 10)[10] than by ideal pathways of safety and security[11] (China Safety Governance Framework, 2024), particularly in regulatory structures that seek to interact with embedded technologies, including modules of Artificial Intelligence (hereafter AI), within neurotecnologies.

Indeed, this is a transnational phenomenon, articulated in different ways, although focused, according to the recent Global Index on Responsible AI (2024, 9), on fostering a concept that can be operationalized on many of these fronts:

> Responsible AI refers to the design, development, deployment, and governance of AI in a way that respects and protects all human rights and upholds the principles of AI ethics through every stage of the AI lifecycle and value chain [...]" requiring that "[...] all actors involved

[8] Os direitos e garantias expressos nesta Constituição não excluem outros decorrentes do regime e dos princípios por ela adotados, ou dos tratados internacionais em que a República Federativa do Brasil seja parte. In english: The rights and guarantees expressed in this Constitution do not exclude others arising from the regime and principles adopted by it, or from international treaties to which the Federative Republic of Brazil is a party.

[9] This is a quote from Peter Häberle on Article 42 of the 1991 Constitution of Estonia, which states: "the rights, liberties and duties listed in the present Chapter shall not preclude other rights, liberties and duties which are in the spirit of the Constitution, or are in concordance with it. A obra mencionada é uma homenagem ao autor, em suas aspirações presentes em seu constitucionalismo cultural, vindas de Goethe, no caso, 'Primal Words. Orphic in Goethe: on the morphology. Tübingen, 1820. "As on the day that lent you to the world the sun stood to greet the planets, you instantly thrived and continued to do so In accordance with the law by which you made your appearance. Thus, you must be, you cannot escape yourself, thus, sibyls and prophets have already spoken, and no passage of time nor any power can break into bits a moulded form that develops as it lives.

[10] "There are those who see in the technology risks at the scale of the Terminator, with the threat of human extinction at the hands of killer robots, bioweapons, or a host of other intende dor unintended innovations. Then, there are those who see it as humanity's saving grace, with the potential to solve a range of human problems, from cancer to climate change. As the capabilities of this technology rapidly become apparent, legislators are placed in the almost-impossible position of trying to preserve the upsides of this emerging technology while avoiding the potentially catastrophic consequences. [...] To some extent, governments are legislating in the dark, as they set the ground rules for technology presenting hypothesized existential risks, some present-day dangers, and potential long-term socially transformative benefits"(G'sell 2024, 10).

[11] This is an interesting distinction proposed in the AI Safety Governance Framework released on 9 September 2024 by China in its Cybersecurity Standardisation Technical Committee. Without going into the merits of the proposal, it is worth remembering that China is spearheading the Global AI Governance Initiative, in which many countries from the global south are at least informally aligned.

in the national AI ecosystem take responsibility for the human, social, and environmental impacts of their decisions.

The World Health Organization (WHO 2023), in turn, has been alerting to the need to establish safe, responsible, and inclusive parameters, especially regarding the use of AI in human health care, investing in the need for broad debate on the risks that may be involved for historically vulnerable groups, being an important thematic soft-law structure for the health sector, and therefore for neurotechnology modalities that incorporate artificial intelligence into their operations or external relations.

Human beings and the development of their personalities are shaped in these spaces, and in the case of the right to health, it emerges as a construction in response to new impacts and risks, both in a negative sense (imagined effects and causalities) and in a positive sense (failing to promote these structures may affect millions of people who could have access to such technologies, improving their well-being, or preventing them from being subjected to sudden and impactful violence, which is relatively foreseeable or capable of being mitigated).

It is possible to emphasize that, in Brazil, the legal and regulatory framework for digital health is a project still under development, and it is in this developing environment, with its partially pre-confirmed foundations, that health-related neurotechnologies are and will be inserted.

In addition to the previously mentioned guidelines, internalized as public policy strategies and constitutional amendments—albeit specific—that are integrated into broader planning, the regulation of so-called Telehealth, under Law No. 14,510/2022, can be highlighted. This law addresses pertinent aspects that merit anticipation in this text, considering the potential relational dynamics with neurotechnologies and their marked intersections with medical practices open to their incorporation, as well as the mechanisms that may be constituted through AI or related technologies of the same kind (in support and technical and practical interoperability).

Telehealth/telemedicine extends far beyond the narrow implications of their terminology. These concepts encompass more than mere video consultations and digital documentation; they represent a comprehensive framework for validating both formal and informal medical practices, adapted to regional variations. This is particularly significant in a country like Brazil, characterized by its vast geographical expanse and diverse healthcare infrastructure and resources, especially in the context of health and medical practice.

For instance, a citizen residing in a rural area without access to specialized medical care might benefit from remote neuroimaging analysis or, in the case of having undergone a brain implant, could have their brain activity monitored remotely. Such advancements not only enhance the likelihood of accurate diagnoses but also reduce the burden of managing specialized healthcare needs, alleviating excessive self-care responsibilities. Depending on the pathology in question, these innovations could either expedite the identification of medical conditions or significantly improve the patient's prospects for recovery and quality of life.

In the telehealth modality, specificities must be considered, whether within the public or private sector. Generally, however, the legislation under discussion authorizes this modality of "providing health services remotely, using information and communication technologies, which involves, among other things, the secure transmission of health data and information through texts, sounds, images, or other appropriate means". The legal endorsement of this type of service provision is reinforced when it is stated that "the acts of health professionals, when practiced through telehealth, shall be valid throughout the national territory" (art. 26-B, caput and sole paragraph, Law 14,510/2022). Among the dogmatic principles governing this normative microcosm, the following are emphasized: the autonomy of health professionals, the patient's free and informed consent, safe assistance, data confidentiality, the promotion of universal access for Brazilians to health actions and services, and digital responsibility (art. 26-A, Law No. 14,510/2022).

Furthermore, the legislation grants health professionals "the freedom and full independence to decide whether or not to use telehealth" (art. 26-C, Law No. 14,510/2022). Additionally, telehealth practice must comply, not only but fundamentally, with the provisions of Laws No. 12,965/2014 (the Brazilian Internet Civil Framework), 12,842/2013 (Medical Act Law), 13,709/2018 (General Data Protection Law), 8,078/1990 (Consumer Protection Code), and, when applicable, Law No. 13,787/2018 (Electronic Health Records Law).

It is noteworthy that the legislation reinforces the responsibility of health professionals for practices performed in the telehealth modality (art. 26-G, I, Law 14,510/2022). Depending on the interpretative approach used to construct the normative framework of these legal provisions, a broad scope for applying neurotechnologies in telehealth service delivery can be discerned. However, it is important to reflect on the provision set out in art. 26-G, I, first part, which stipulates that telehealth must be "performed with the free and informed consent of the patient or their legal representative," given the potential serious externalities that neurotechnologies may reveal, depending on the treatment, particularly regarding the legal implications for the free development of personality, connected with the protective legal sphere of self-determination as a patient.

In particular, with respect to acts considered, within the Brazilian legal system, as exclusive to medical professionals, it is worth reflecting on the levels of assistance, support, or even exclusive decisions—or "micro-decisions"—in which digital technologies, or in this case neurotechnologies, may be involved, requiring necessary normative accommodations, to the relationship model and the type of neurotechnology used, and their respective contextual risks and benefits (risk-benefit approach).

In Law No. 12,842/2013, which regulates these rapidly evolving spheres of action, it is reinforced that "the object of the physician's practice is the health of human beings and human communities," for whose benefit "the physician shall act with utmost diligence, to the best of their professional ability, and without discrimination of any nature" (Article 2, *caput*). The physician shall develop their professional actions in (i) the promotion, protection, and recovery of health; (ii) the prevention, diagnosis, and treatment of diseases; and (iii) the rehabilitation of the infirm and individuals with disabilities. The aforementioned legal base strengthens the responsibility of this professional by indicating as exclusive activities: the indication and execution of surgical interventions and prescription of pre- and post-operative medical care, the indication of invasive procedures, whether diagnostic, therapeutic, or aesthetic, and, without exhausting the list, the determination of prognosis related to medical diagnosis (Article 4, I, II, X).

Reflecting on the 'ecosystem' of health digitalization and medical practices, it is noted that, beyond the state structures capable of acting as supervisors of medical processes (from direct and indirect public administration), Brazil has a microsystem (as occurs in other countries), in which the Federal Council of Medicine and the Regional Councils[12] are delegated the issuance of acts and procedural actions involving, especially, the medical activities and practices in question.

In the Code of Medical Ethics, a deontology related to technoscientific practices is projected, stating that "in applying knowledge created by new technologies, considering their repercussions on both present and future generations, the physician shall ensure that individuals are not discriminated against for any reason linked to genetic heritage, protecting their dignity, identity, and integrity, while aiming to couple medical practice with the exercise of available technical and scientific means that seek the best results" (CFM, 2018, Chapter I, Fundamental Principles, XXV, XXVI, CFM Resolution No. 2,217).

Another structural point is the legal incentive, which maintains an architectural coherence with its 'extralegal effects,' in the possible digitalization of medical identity (Medical Identity Cards—CIM, for E-CRM—Digital Medical Identity Card), having as its primary purpose, clearly, the identification of the professional (a myriad of personal data), but also the pursuit of operationalization in digital systems, creating legal and technological frameworks in which performed acts can be securely linked to the practicing individual, and also to prevent fraud or identity theft capable of causing harm to third parties and to the professional themselves. This matter is of particular importance in terms of creating information structures concerned with the cybersecurity of systems directly or indirectly linked to neurotechnologies in the medical field.

It is determined, for instance, that the E-CRM encompasses:

> A set of updated information, organized and encoded in computer media, which may [...] serve as an important instrument to establish a secure channel for commuting information between human and technological means, allowing for the verification and prevention of certain individuals using false documents to impersonate a physician.

In conjunction with this identification system, the Medical Credential application is created, wherein acts are subject to electronic authentication, with the secondary

[12] It can be observed that, in certain Regional Councils, one can view "opinions" stating the inexistence of "robust scientific evidence for the Therapeutic Indication of Transcranial Direct Current Stimulation for the Treatment of Autism Spectrum Disorder in Medical Assistance, which may be conducted as Medical Research, after approval of a specific protocol by the Research Ethics Committee of the pertinent Institution.

purpose, as mentioned, of "regulating the use of technologies in the medical field to identify professionals in a secure and integral manner".

Returning to a brief illustration of the Brazilian regulatory framework linked to the digitalization of health and medical practices, it is important to consider that, in the case of neurotechnology, many of their applications are still in research and testing stages, even though one can observe, compared to a few years ago, a widespread diffusion of this technological modality's potential in academic and clinical environments, but also in the diffuse culture of society. This issue, which will not be addressed in the present study, possesses global dimensions and is frequently confined to the scope of regulatory law, encompassing the duality of health and consumer applications, with the regulatory order for research linked to its background. The intersections with consumer goods associated with the wellness industry are often ambiguous and, at times, problematic. These goods influence, capture, and induce behaviors related to self-tracking of health through devices that are typically self-regulated within private-sector frameworks. The impacts of such devices on physical and psychological health remain largely unclear, compounded by the additional concern of their use without oversight from qualified health professionals.

The Brazilian neuroscientist, Nicolelis (2001, 2024), with the publication in 1999 in the journal *Nature Neuroscience*, and with the subsequent text of 2001, entitled Actions from thought, in which he coined the terminology *brain-machine interfaces* (BMIs), employed in health and general regulations worldwide, and by the neuro industry itself, especially those concerned with Artificial Intelligence, beyond his scientific works, e.g., recently launched the science fiction book Nothing Will Be the Same Again, in which he explores some challenges in an open and imaginative manner, contributing to and in the (critical) insertion of this form of technology in society. It is impossible not to reference the contribution—whether positive or negative—of actor Elon Musk, with the company Neuralink—and also through his actions in the media—to the accelerated diffusion and widespread 'cultural acceptance' of these technologies.

To illustrate that, from scientific and academic activities to the industry itself, growth is imprinting, not only as a reaction to or reason for a considerable increase in psychophysical pathologies related to the brain worldwide, but also as a means to capitalize on new relational forms that the brain and computer (and bio or pharmacological neurotechnologies) can create or remodel, such as in games, wearables, and brain-enhancement devices. Thus, beyond the health system, however, connected to it, there is an exploration of new market spaces in which neurotechnologies present themselves as central.

In this polycontextual scenario, the Brazilian state recently presented new legislation related to research with human beings, which corresponds to a specific and perhaps the most important phase in the development of neurotechnologies.

The Brazilian Law No. 14,874 of May 28, 2024, which provides for research with human beings and establishes the National System of Ethics in Research with Human Beings in Brazil, is applicable to public and private institutions (Article 1, *caput*, Law No. 14,874/2024).

This legislation, largely considered a setback in the protection of individuals and therefore being contested regarding its constitutionality before the Brazilian Constitutional Court, enhances the governance structure applicable in the area under discussion. Namely, it proposes a close relationship with all health authorities in Brazil (the entity competent to regulate, control, and supervise products and services that involve risk to public health), through an institutional process that fosters dialogue with other public and private entities, which will have to establish a Research Ethics Committee (CEP), and thus related to the national instance of research ethics, being an interdisciplinary and independent collegiate body, integral to the Ministry of Health (federal sphere), under the coordination of the technical area responsible for the field of science and technology, of a *normative, consultative, deliberative, educational nature, competent to proceed with the regulation, supervision, and ethical control of research, with a view to protecting the integrity and dignity of research participants, and to contribute to the development of research within ethical standards* (Article 2, subsections X, XXXVI, Law 14,874/2024).

Moreover, in addition to other frameworks that could be cited, reference is made to the document on Good Clinical Practices (GCPs), which will serve as a veritable constitution for this subsystem. It is understood as the:

Standard defined in regulation, in accordance with international norms and best practices, for the planning, conduct, execution, monitoring, auditing, recording, analysis, and reporting of research, with the aim of ensuring the credibility and validity of data and results, as well as protecting the rights, integrity, and confidentiality of research participants' identities (art. 2, item VII, Law 14,874/2024).

The need to establish prior (bio)ethical reviews for any research involving human subjects is reinforced, in order to guarantee ethical procedures, dignity, well-being, and the safety of participants (art. 6, Law 14,874/2024). The conceptual framework outlined by this legislation also proposes complementary distinctions, specifically concerning scientific, technological, or innovation research involving human[13] subjects, clinical research involving human subjects,[14] research with human subjects,[15] and the concept of multicentric research, which is conducted in different study centers by more than one researcher, while adhering to a single protocol (art. 2, XXXV, Law 14,874/2024) (potentially allowing for transnational collaborations).

---

[13] Scientific, technological or innovation research involving human beings: research that, individually or collectively, interacts directly with human beings, without the purpose of registering the product under research.

[14] XXXIII—clinical research with human beings: a set of scientific procedures systematically developed with human beings with a view to: (a) evaluating the action, safety and efficacy of medicines, products, techniques, procedures, medical devices or health care, for therapeutic, preventive or diagnostic purposes;

(b) to verify the distribution of risk factors, diseases or illnesses in the population; (c) evaluating the effects of factors or states on health;

[15] XXXIV—research with human beings: research that, individually or collectively, involves human beings, in whole or in part, and involves them directly or indirectly, including the handling of their data, information or biological material;

These regulatory considerations are closely tied to the processes of conceptualizing and developing neurotechnologies, particularly during the research and clinical trial phase, and the legal statute commented on anticipates a finalistic criterion, when research is aimed at the health and medical areas it is explicitly stated in this legal framework that the definition of a medical device includes any "instrument, apparatus, equipment, implant, in vitro diagnostic device, software, material, or other article intended by the manufacturer to be used, either alone or in combination, in humans […]" (art. 2, item XVII, Law 14,874/2024).

The opening towards the international landscape is incorporated, notably through the established document on Good Clinical Practices (a standard defined in regulation, following international norms and best practices […]). However, this is not entirely novel within the Brazilian regulatory system, considering the existence—though not in identical terms—of Resolution RDC No. 346, dated March 12, 2020, from the National Health Surveillance Agency (Anvisa, 2020), which defines criteria and extraordinary and temporary procedures for the certification of good manufacturing practices for the purposes of registering and post-registration changes to active pharmaceutical ingredients, medicines, and health products (in response to the international public health emergency of the new Coronavirus) (art. 3, sole paragraph, items I, II, and III, Resolution No. 346/2020).

The regulatory ecosystem governing the research and development of neurotechnologies for healthcare and medical practice is also structured around Resolution 751 of 2022, issued by the National Health Surveillance Agency (Anvisa). This agency is responsible for establishing guidelines and overseeing regulations related to risk classification, requirements, labeling, and usage instructions for medical devices, authorising or not its commercialisation and use in the country.

However, neurotechnological applications targeting the consumer market are not strictly governed by these limited regulatory frameworks, or those consumer goods that are usually classified as self-track technologies, such as smartwatches or other similar devices for sport, fitness, yoga, concentration, self-health and so on, which create a blurred boundary between medical and wellness equipment.

Instead, they are subject to protective provisions within the consumer protection legal system, as is commonly observed in most national jurisdictions and at the international level, with few *ex ante* or *ex post* regulatory duties, as well as with adherence and validation in social and cultural uses and practices that are sometimes incorporated as harmless. The new neurotechnologies that have been developed and are being developed will be part of this preconceived universe.

It is important to recover the right to health, to data protection, to the development of the personality and to the dignity of the human person, also to make this lawless region reflect on its risks, especially with the increased reach of state surveillance and large private companies. The emerging grammar of neurorights suggests and aims to fulfill this regenerative purpose of Law, for the neurotechnological environment.

To analyze this scenario, including its deficiencies and potential contradictions, it is essential to consider another dimension of law—namely, Human and Fundamental Rights Law, with particular attention to the emerging concept of neuro-rights. In the next section, an analysis will be conducted on the general framework for regulating artificial intelligence in Brazil, with particular attention to its nuances concerning the topic at hand. This analysis will engage in a dialogue with the concept conventionally referred to as neurorights, situated within the broader dimension of human and fundamental rights. Furthermore, it will examine Brazil's specific proposal to elaborate on this rights framework, exploring its technoscientific connections to the human nervous system and its intersections with other regulatory domains and contextual legal-social dynamics.

4 Neuro-rights and Neurotechnologies: A Necessary Self-constitutive Dialogue Between Regulatory Law and Human and Fundamental Rights Law for New Digital Spaces

In light of the above, a normative gap is perceptible, although it is possible to capture the gradual evolution of new meanings and guidelines, which are and will be perpetuated in disputes over economic, social, medical, scientific, political and therefore legal interests. It is reasonable to identify this scenario, even though there are numerous administrative and constitutional concerns involving public and private sectors, which can accommodate the challenges that neurotechnologies pose in the global society, particularly in Brazil, notably within the constitutional health system. There is, in fact, a developing legal-normative regime and health, research, and medical practices that can be regarded as more advanced, at least within their own environments and organizations, although limited when viewed through a broader analysis of the country's regulatory framework, being this particularly evident in the lack of guidelines for the consumer market, not to mention the potential applications for educational and military purposes or in criminal prosecution contexts, which require specific and meticulous attention.

The legal framework within Human and Fundamental Rights Law, which is under development, particularly related to neuro-rights, should serve as a support tool in this context, in order to increase attention to key points, closely aligning with AI regulatory initiatives. This alignment is essential at this stage of technological development in the technoscientific field, precisely because of the continuing complex digitalisation of neurotechnological practices, and even with AI mechanisms embedded in such applications, be they internal or external material devices, or so-called micro or macro biocomputers.

In this regard, it is important to highlight the potential of cerebral capital in digital terms and its multiple implications. To do this, we will firstly look at the general regulatory structure that is being conceived in Brazil, related to Artificial Intelligence, subsequently placing it in addition to the legal thoughts related to what has been called neuro-rights, particularly in Brazil and in Latin America, illustrating these matters with the specific regulatory proposal, still embryonic, although nascent, which has the purpose of infra-constitutionally protecting the brain and the central nervous system of individuals, in a comprehensive way. This will serve to paint a picture of the emergence of state law, in the language of human and fundamental rights, and of regulatory law, which can be characterised as fragmented, revealing that, in addition to the scientific and technological complexity that the subject raises, the area of law still requires greater care and reflection to be developed, taking into account the existing normative preconditions, those in the process of being idealised and, therefore, the need for evolutionary and sectoral adaptation, but also having an inserted self-reflexivity, in order to be clear about the thematic scope and the field that is being co-normatised.

Referring to the Global Index on Responsible AI (from November 1, 2021, to November 1, 2023), Brazil, out of 138 countries assessed, was ranked 18th, with the highest scores given to initiatives carried out by "non-state actors." In the section on "responsible AI capacities," however, Brazil received the lowest score among the six evaluated categories. Among the "bright spots" in Brazil's structure is the Algorithmic Impact Analysis (AIA), anticipated in the Brazilian Artificial Intelligence Strategy, and foreseen in Bill No. 2338 of 2023 (currently under review in the Brazilian parliament), which aims to become the civil framework for AI in the country. Regarding Algorithmic Impact Assessment (AIA), highlighted in the aforementioned Global Index, it is an extremely relevant instrument, both in terms of protecting rights and for determining the risk level of a given AI system (and its consequent regulatory obligations). It is also important for the computing and AI application industry, where, depending on the economic sector in which it operates, more obligations may apply compared to others. The AIA includes a framework for dialogue specifically tailored to the computational modality of artificial intelligence. This framework can extend to specialized branches of application, particularly neurotechnologies that integrate AI. Moreover, it encompasses the sector-specific particularities associated with these technologies, which are still in the process of development in Brazil, as previously mentioned.

In Brazil, the Bill No. 2338/2023, in its latest version published on 07/04/2024, aims to establish general national application rules affecting the design, development, implementation, use, adoption, and responsible governance of AI systems, with the purpose of protecting fundamental rights, promoting responsible innovation, and ensuring the implementation of safe and reliable systems for the benefit of human beings, the democratic regime, and social, scientific, technological, and economic development (art. 1). In the evolving Brazilian regulatory framework, there will be an approach based on risk and rights. Among the anticipated rights are the rights to information, privacy and data protection, human participation and oversight, and the right to non-discrimination and protection from abusive practices. Additionally, for AI systems aimed at vulnerable groups, there will be a mandate to follow a transparency procedure throughout the entire lifecycle, using simple, clear, and age-appropriate language, considering cognitive capacity (art. 5, I, II, III, IV, § 2).

This regime also includes rights related to individuals or groups affected by AI systems that produce significant legal or high-risk effects, expanding the right to an explanation of decisions, the right to contest and request a review of decisions, and the right to human review of decisions, taking into account context, risk, and the state of the art of the technology, with guidelines to be set forth in regulations (art. 6, I, II, III, sole paragraph). Notably, the competent authority (the coordinating authority), in this case, the National Data Protection Authority (ANPD), as part of the National System for AI Regulation and Governance (SAI), will act jointly in formal institutional cooperation with sectoral authorities to establish guidelines on how and under which conditions individuals may exercise their rights in relation to each AI system operator.

Regarding the "Algorithmic Impact Assessment (AIA)," this is defined in the regulation project as an analysis of the impact on fundamental rights, presenting preventive, mitigating, and corrective measures for negative impacts, as well as measures to enhance the positive impacts of an AI system (art. 4, XVI), which must be carried out for high-risk AI systems.

It is reasonable to foresee its application in the health sector, given the trend towards applications involving human research and medical practices. Specifically, this trend is reflected in the text, which identifies high-risk AI systems as those developed and used for purposes and contexts that take into account the probability and severity of adverse impacts on individuals or groups, such as applications in the health sector that assist with diagnoses and medical procedures when there is *significant risk to the physical and mental integrity of individuals* (art. 14, VIII). For AI systems involved in organizational administrative processes, the project supports high-risk classification for systems used for evaluating and classifying calls and determining priorities for essential public services, such as *healthcare* (art. 14, V).

The proposed project seeks to align with Law No. 13.709/2018 (LGPD—personal data protection), allowing AI operators required to prepare a data protection impact report to conduct this in conjunction with the AIA (art. 28). It is important to note that this assessment should be conducted as a "continuous interactive process" throughout the entire lifecycle of high-risk AI systems, with periodic updates (art. 27).

The project also establishes a potential relationship between the AIA and Codes of Conduct, allowing operators to develop procedures based on the guidelines outlined in the AIA for high-risk AI applications, and sectoral authorities will have the responsibility to approve these codes within their areas of competence as determined by law (art. 39, § 1, § 2, 'c', § 4, I).

Regarding the regulatory-governance structure, Section I of the National System for AI Regulation and Governance (SAI) sets out an integrative administrative-constitutional approach involving state authorities with the participation of civil society. For example, the SAI will include the National Data Protection Authority (ANPD), which will serve as the competent coordinating authority (art. 44, §, I). Additionally, the participation of sectoral authorities (art. 44, § 1, II), the Permanent Council for AI Regulatory Cooperation (CRIA) (art. 44, § 1, III), and the Committee of AI Experts and Scientists (CECIA) (art. 44, § 1, IV) is encouraged, with the possibility for a subsequent legal act from the Federal Executive Power to define the list of bodies and entities that will be part of the SAI (art. 44, § 2).

It is worth highlighting that the coordinating authority of the SAI, in this case, the ANPD, will have the authority to issue, in collaboration with other SAI members, binding regulations, such as those on procedures and requirements for preparing the algorithmic impact assessment (art. 45, II, ‘b’), as well as normative guidelines related to certification and accreditation of certifying bodies to encourage and ensure better governance practices throughout the AI system lifecycle (art. 45, VIII), and to promote the adoption of internationally recognized standards, best practices, and certifications (art. 45, IX).

In this context, dialogue with sectoral authorities is prescribed, which will be responsible for issuing specific rules for AI applications, including aspects related to high-risk activities (art. 46, II), and for promoting and encouraging regulatory dynamics surrounding Good Practices and Governance (art. 46, III), encouraging the adoption of internationally recognized standards and certification best practices (art. 46, IV), and establishing governance measures tailored to each high-risk AI system application and use to promote harmonization of national legislation and international standards (art. 46, V, ‘a’). To reinforce concerns over high-risk AI systems, the SAI’s coordinating authority, the ANPD, is authorized to conduct investigations into high-risk AI systems in conjunction with sectoral authorities when there is suspicion of violations of the parameters established by the Law or sectoral regulations (art. 48, § 2).

The initiatives surrounding the development of new grammar within Human and Fundamental Rights Law, particularly neuro-rights, are pertinent and coherently aligned with this technological and regulatory context, especially with the increasingly pervasive presence of AI systems, in this case, involving neuro-instruments.

Even though certain voices have been advocating for this legal-protective dynamic since at least the 1990s and early 2000s (Boire, 1999; Sententia, 2004), there has been a parallel growth and maturation of this new language, also as a response to concrete technoscientific advances in recent years and the social transformation currently underway with the datafication[16]/digitalization of life (Mejias & Couldry, 2019).

Many of these advances and potential rights to be included in constitutional texts and international treaties are not simply arising from “popular demands,” as has generally characterized the history of human and fundamental rights, but rather from advocacy by researchers from various fields of knowledge, including legal scholars.

Of particular significance, for instance, is the text by Ienca and Adorno (2017), *A New Category of Human Rights: Neurorights*, which represents a discursive advance focused on the new contours of health technologies linked to human rights concerns. The authors assert that the traditional legal focuses “[...] were not sufficient to protect us from neurotechnology,” thus proposing neuro-rights, comprising the components of “cognitive liberty, mental privacy, mental integrity, and psychological continuity”.

Another prominent example, although not exhaustive, is the work of the Neurorights Foundation, led by Rafael Yuste, which advocates for the incorporation of neuro-rights into the Universal Declaration of Human Rights and calls for the urgent

[16] For a critical perspective: Datafication is not just the making of information, which, in one sense, human beings have been doing since the creation of symbols and writing. [...] datafication is a contemporary phenomenon which refer to the quantification of human life through digital information.

need for national and transnational governments to establish new legal frameworks to prevent potential or actual abuses in the sphere of neurotechnologies. The Neurorights Foundation's (2021) proposal establishes five key guidelines, which constitute the so-called neuro-rights:

> The right to personal identity, as limits should be developed to prohibit technology from altering an individual's concept of self. That is, by connecting human brains to computers, neurotechnologies may blur the line between a person's consciousness and external technological inputs. The right to free will, as people must be able to make autonomous decisions and retain control over the decision-making process and its consequences, being protected from manipulation by unknown external neurotechnologies. The right to mental privacy, which is tied to a notion of autonomy, or the ability for individuals to keep their thoughts protected from arbitrary intervention and disclosure. The right to equitable access to mental enhancement tools, ensuring that the benefits of potential improvements to cognitive and mental abilities resulting from neurotechnologies are distributed fairly. The right to protection against algorithmic distortion/discrimination, ensuring that technology is not designed or operated within a biased environment or that it does not introduce discriminatory elements (Yuste et al., 2021, 160–161).

It is important to mention the significant role of Latin America in the discussion on neuro-rights. As an illustrative example, Chile (2021) was the first country to incorporate neuro-related concerns into its constitutional text. It would be possible and appropriate to mention other Latin American countries[17] and regional initiatives,[18] which, in turn, are influencing other continents and countries, and the international[19] landscape as a whole.

However, particularly considering the Brazilian scenario, initial attention has been focused on the environment of the Federal Constitution of 1988 in the recent post-pandemic years. With the Proposal for Constitutional Amendment 29 of 2023 (PEC 29/23) (Federal Senate, 2023), there is an aim to prioritize in the constitutional text, among fundamental rights, the protection of mental integrity and algorithmic transparency. As mentioned at the beginning of this text, Brazil possesses a dense federalized constitutional structure, and this is reflected in the matter under discussion, given that there exists, at the subnational level, specifically in the Constitution of the State of Rio Grande do Sul (1989), the recent approval of Amendment Proposal No.

---

[17] As an illustration: Uruguay, Argentina, Colombia.

[18] Organização Dos Estados Americanos. Declaração do Comitê Jurídico Interamericano de Neurociência, Neurotecnologias e Direitos Humanos: Novos Desafios Jurídicos para as Américas.

Organização Dos Estados Americanos. Comitê Jurídico Interamericano. Declaración de princípios interamericanos en materia de neurociencias, neurotecnologías y derechos humanos.

Parlatino. Ley Modelo de Neuroderechos Para América Latina y el Caribe.

[19] European Parliament. Panel for the future of science and technology (STOA)—The protection of mental privacy in the area of neuroscience—societal, legal and ethical challenges.

Council Of Europe. Common Human Rights Challenges Raised by Different Applications of Neurotechnologies in the Biomedical Field.

European Parliament. Artificial intelligence in a digital age. Resolution of 3 May 2022.

United Nations Human Rights Council. Neurotechnology and human rights, 2023.

Unesco. Report of the International Bioethics Committee of UNESCO (IBC) on the ethical issues of neurotechnology. 2021.

OECD. Recommendation on Responsible Innovation in Neurotechnology. 2019.

298/2023, which established the protection of mental integrity as a guideline of the state's (constitutional) science and technology policy, stating that it "shall be based on respect for life, health, human dignity, the *mental integrity of human beings*, and the cultural values of the people [...]" (art. 235, caput, sole paragraph).

Briefly addressing another legal angle, reference is made to infraconstitutional initiatives at the federal level, with Bill 522/2022 (and another) (2022), aimed at modifying the aforementioned General Law for the Protection of Personal Data (LGPD), to include, conceptualize, and regulate the protection of so-called neural data. This is a normative scope forged from the dogmatics of personal data protection, being of special relevance to and with the relationships of 'datafied' neurotechnologies, and thus in the development of AI systems designed for the field.

Complementarily, in the draft bill for the reform of the Brazilian Civil Code presented by the Commission of Jurists (2024), there is a parallel modification to the LGPD, so as to include in art. 2 of the aforementioned legislation, "the discipline of personal data protection has as its foundations: VIII—mental privacy, cognitive freedom, and mental integrity," with coherential importance, the proposal for addition, in this case, specific to the Civil Code, with art. 17, whose content safeguards that "every person has the right to the recognition and preservation of their personal identity, composed of the set of attributes, characteristics, behaviors, and choices that distinguish them from others".

This infraconstitutional initiative, seemingly modest, has a protective potential that complements the aforementioned measures. Additionally, one may reference Bill No. 2.174 of 2023, which aims to create a general and comprehensive regulatory framework capable of encompassing "all relations" in which the brain and human nervous system are, in any way, affected. The summary of the bill states: "it establishes the norms and principles for the protection of fundamental rights related to the brain and the human nervous system, aiming to ensure the protection and promotion of individuals' neuro-rights".

The proposed language establishes the layers constituting neuro-rights, affirming that "neuro-rights are fundamental rights related to the brain and the human nervous system, including, but not limited to: (i) the right to cerebral and neurological integrity, (ii) the right to cerebral and neurological privacy, (iii) the right to cognitive freedom, (iv) the right to cognitive equality, (v) the right to education and neuroscientific information, (vi) the right to personal autonomy and free will, and (vii) the right to non-discrimination based on neurological characteristics (Article 2, caput, items).

This is a non-exhaustive list of open and flexible terms, adaptable to various neurotechnological contexts. Article 3 introduces the "State's duty" to ensure the protection of neuro-rights and to promote measures for the prevention and combat of violations of these rights. It is important to note that, along the legislative path, the goal is to normatively solidify each of these rights.

For example, it is stated that the right to cerebral and neurological integrity pertains to the protection against any form of intervention or forced modification of the brain or human nervous system, including the right to reparations in case of neurological harm (Article 5). Regarding the right to cerebral and neurological privacy, it relates to the protection against the unauthorized collection, storage, processing, sharing, or use of brain or neurological information (Article 6). Moreover, the right to cognitive freedom is connected to the right to freely think, imagine, create, and express ideas, concepts, emotions, and feelings, without censorship or coercion (Article 7). Additionally, the evolving doctrine on the subject addresses cognitive equality, understood as the right of all people to have equal opportunities for cerebral and neurological development, free from any form of discrimination (Article 8).

A brief overview of the legal framework reveals the establishment of a comprehensive legal provision that grants every individual the right to exercise autonomy over their own brain and nervous system, relating this to new dimensions of physical and mental integrity. The use of brain modification techniques without the individual's free, informed, and explicit consent is prohibited, except in cases of essential medical treatments for the preservation of life or health, provided that ethical and legal requirements are met (Article 14). Furthermore, the use of mind-reading techniques without consent is prohibited (Article 15), though the right to privacy of brain activities is ensured, forbidding monitoring without authorization or a court order (Article 16—controversial provision). Article 17 reinforces the prohibition of the use of brain persuasion or manipulation techniques without the individual's free, informed, and explicit consent.

Regarding health and human research, it encompasses, for instance, the duty for all research involving brain modulation techniques to be conducted in accordance with ethical and legal principles, requiring the explicit and written consent of the individual (Article 18). Additionally, the right of individuals to access and control their own brain information is established, prohibiting the use of such information without authorization or a court order (Article 19).

The referenced legislation establishes a general framework for neuro-rights, which can be extended to sector-specific contexts, including the integration of rights across both analogue and digital domains. These sectors encompass research, medical practice, consumer relations, state intelligence operations, and criminal prosecution. The law reflects the Brazilian state's interest in developing mechanisms for the exercise of rights, fostering their incorporation into sectoral regulatory frameworks, and ensuring their recognition by the judiciary. Notably, it also addresses modalities of "direct intervention" in the central and peripheral nervous systems, which can be performed with judicial authorization, for investigative purposes and other forms of state-authorized violence (an extremely problematic legal provision from a human rights perspective).

This cross-fertilization of legal regimes, whether from the law of human and fundamental rights, or from regulatory law, poses the problematic situation to be faced: there is no materialized regulatory structure on the subject, even if imperfect. There are, as explained, spaces configured under construction, although possibilities for co-productive dialogues. Brazilian state law, both constitutional and infraconstitutional, are undergoing profound transformation, particularly in the health and medical sector, driven by inevitable intersections with research and clinical operations. New practices, infrastructures, and relationships are emerging, heralding both subtle and revolutionary changes. The field of neurotechnologies has undoubtedly been the focus of considerable attention within this new global design in which Brazil aims to be a leading player, definitively overcoming the constraints of imposed technocolonialism. This text seeks to briefly present the related initiatives currently in dialogue, within a normative network still in the process of maturation.

5 Conclusion

Recently, at the Stanford HAI Healthcare AI Policy Workshop (2024), an informal vote was conducted among the participants, asking how "effective governance of healthcare AI can be achieved." Regarding regulation, the option "minor changes to existing regulation" received the fewest votes, with less than 10%. The alternative "substantial changes to existing regulation" was the most voted, garnering nearly 60%. The proposal for "a novel regulatory framework" was the second most voted, with almost 40% of votes.

This event, held in the United States of America (USA), is mentioned here to illustrate that, regardless of the context in question, the demand for a regulatory structure related to AI applications in healthcare is a growing call from the sector and its stakeholders, given the multifaceted nature of these initiatives and, especially, the risks posed to the individuals targeted by such applications.

The terms, sectoral dialogues, obligatory definitions, and thus a reasonable coherence among public and private institutions, both national and international, need to be conceived for this 'new digital era.' Neurotechnologies, in its biochemical and computational facets, with the incorporation of artificial intelligence, or with an external relationship with this technology, in particular, present unique challenges. The responses must be carefully crafted, aiming, from a legal-normative perspective, to accommodate the concrete protection of human and fundamental rights. In Brazil, there is an ongoing process in the field of 'digitalized' health, influenced by constitutional and infraconstitutional initiatives applicable to both the public and private sectors.

Neuro-rights, alongside these regulatory initiatives, notably tied to AI applications, are byproducts of a new evolutionary stage of modern society. In the Brazilian constitutional system, and its infraconstitutional developments geared toward these ends, initiatives have been launched, albeit still modest and, depending on the perspective, insufficient. The general regulatory framework for AI, currently under development in Brazil, is an important step in this scenario. Sectoral supplements will inevitably be necessary, which the project itself foresees in its regulatory program. However, sectoral legal provisions involving health, medical practice and research will also need to be updated, especially for the innovations in view in the neurotechnological field, notably with those technical devices made with embedded or non-embedded AI. These necessary transformations and adaptations reflect, in fact, a process of consolidation of the sociological analysis of a computational ubiquity. The fundamental right to health, in the Brazilian constitutional subsystem, still remains with its normativity uncertain in this space under construction. However, its normative force will be exact, to be established together with the other rights to be enshrined for this neuro era.

This work was conceived with the purpose of conducting a brief mapping of this scenario to facilitate its understanding, even in its generalities. This is, at first glance, a preliminary step toward suggesting new normative and constructive approaches to operate within this increasingly complex world.

References

Albers, M. (2016). A complexidade da proteção de dados. *Direitos Fundamentais & Justiça, 35*, 19–45.

Anvisa. (2020). *Resolution RDC No. 346, of March 12, 2020*, Brasília.

Anvisa. (2022). *Resolution RDC No. 751, of September 15, 2022*, Brasília.

Ascensão, J. O. (2008). A dignidade da pessoa e o fundamento dos direitos humanos. *Revista da Faculdade de Direito da Universidade de São Paulo, 103*, 277–299.

Bill No. 522, of 2022. (2022). *In order to conceptualize neural data and regulate its protection*, Brasília.

Bill No. 2.174, of 2023. (2023). *Estabelece as normas e princípios para proteção dos direitos fundamentais relacionados ao cérebro e ao sistema nervoso humano, objetivando garantir a proteção e promoção dos neurodireitos dos indivíduos*, Brasília.

Bill No. 2338, of 2023. (2023). *Dispõe sobre o uso da Inteligência Artificial*, Brasília.

Boire, R. G. (1999). On cognitive liberty: Part I. *Center for Cognitive Liberty and Ethics, 1*(1), 7–13.

Brazilian Strategy for Digital Transformation (E-Digital 2022–2026). (2022). *2022–2026 Cycle*, Brasília.

CFM. (2018). *Resolution No. 2. 217 of 2018.*

Chile. (2021). Ley 21383. Modifica La Carta Fundamental, para estabelecer el desarrollo científico y tecnológico al servicio de las personas, 2021.

China Safety Governance Framework. (2024). 9 September, 2024.

Commission of Jurists. (2024). *Comissão de Juristas responsável pela revisão e atualização do Código Civil*, Brasília.

Constituição da República Federativa do Brasil de 1988. (1988). *Constituição da República Federativa do Brasil*, Brasília, 1988. Recuperado em: 10 set., 2024, de https://www.planalto.gov.br/ccivil_03/constituicao/constituicao.htm

Constitution of the State of Rio Grande do Sul of October 3, 1989. (1989). *Constituição do Estado do Rio Grande do Sul*, Porto Alegre.

Costa, J. (2023). *Health as a social system: Luhmann's theory applied to health systems—An introduction.* Majuskel Medienproduktion.

Digital Health Strategy for Brazil (2020–2028). (2020). 2020–2028 Cycle, Brasília.

FDA. (2017). *FDA approves pill with sensor that digitally tracks if patients have ingested their medication.* Recuperado em: 10 set., 2024, de https://www.fda.gov/news-events/press-announcements/fda-approves-pill-sensor-digitally-tracks-if-patients-have-ingested-their-medication

Federal Senate. Proposal for Constitutional Amendment No. 29, of 2023. (2023). *Amends the Federal Constitution to include, among fundamental rights and guarantees, the protection of mental integrity and algorithmic transparency*, Brasília.

G'sell, F. (2024). *Regulating under uncertainty: Governance Options for generative AI.* Stanford Cyber Policy Center.

Global Index on Responsible AI. (2024). South Africa: Global Center on AI Governance.

Goethe in the Thought of Peter Häberle. (2024). *Classical Literature in Constitutional Evolution.* Anna Amalia und Goethe Stiftung.
Hanemaayer, A. (2021). Don't touch my stuff: Historicising resistance to AI and algorithmic computer technologies in medicine. *Interdisciplinary Science Reviews, 46*, 126–137.
Hoffmann-Riem, W. (2021). *Teoria geral do direito digital: Transformação digital desafios para o direito.* Forense.
ICT in Health 2023. (2024). *ICT in Health: survey on the use of information and communication technologies in Brazilian healthcare facilities.* Comitê Gestor da Internet do Brasil.
Ienca, M., & Adorno, R. (2017). *A new category of human rights: Neurorights.*
Karaboga, M., Matzner, T., Obersteller, H., & Ochs, C. (2016). Is there a right to offline alternatives in a digital world? In R. Leenes, et al. (Eds.), *Data protection and privacy: (In)visibilities and infrastructures.* Springer.
Kirste, S. (2016). Dignidade humana e direitos humanos – ontologia ou construtivismo dos direitos humanos – menchenwürde und menchenrechte – ontologie oder konstruktivismus der menchenrechte. *Hendu, 7*, 2–15.
Law No. 8.080, of 19th September of 1990. (1990). *Dispõe sobre as condições para a promoção, proteção e recuperação da saúde, a organização e o funcionamento dos serviços correspondentes e dá outras providências*, Brasília.
Law No. 12.842, of July 10, 2013. (2013). *Dispõe sobre o exercício da Medicina*, Brasília.
Law No. 13.709, of August 14, 2018. (2018). *Lei Geral de Proteção de Dados Pessoais (LGPD)*, Brasília.
Law No. 14.510/2022, of December 27, 2022. (2022). *Autoriza e disciplina a prática de telessaúde em todo o território nacional*, Brasília.
Law No. 14.874, of May 28, 2024. (2024). *Dispõe sobre a pesquisa com seres humanos e institui o Sistema Nacional de Ética em Pesquisa com Seres Humanos*, Brasília.
Mejias, U. A., & Couldry, N. (2019). Datafication. *Internet Policy Review, 8*(4), 1–10.
National Artificial Intelligence Plan (2024–2028). (2024). Cycle 2024–2028, Brasília.
National Digital Government Strategy (2024–2027). (2024). Cycle 2024–2027, Brasília.
NeuroRights Initiative. (2021). *The five ethical NeuroRights.*
Nicolelis, M. (2024). *Nada mais será como antes.* Planeta Minotauro.
Nicolelis, M. (2001). Actions from thoughts. *Nature, 409*, 403–407.
OECD. (2019). *Recommendation of the Council on responsible innovation in Neurotechnology.*
Sarlet, I. W. (2015). *Dignidade da pessoa e direitos fundamentais na Constituição Federal de 1988.* Livraria do Advogado.
Sarlet, G. B. S., & Ruaro, R. L. (2021). A proteção de dados sensíveis no sistema normativo brasileiro sob enfoque da Lei Geral de Proteção de Dados (LGPD – L. 13.709/2018). *Rev. Direitos fundam. Democ., 26*(2), 81–106.
Sententia, W. (2004). Neuroethical considerations: Cognitive liberty and converging technologies for improving human cognition. *Annals of the New York Academy of Sciences, 1013*, 221–228.
Spiecker gen. Döhmann, I. (2023). The more the merrier: A dynamic approach learning from prior misgovernance in EU data protection law. In M. C. Kettemann, A. Peukert, I. Spiecker gen. Döhmann (Eds.), *The law of global digitality.* Routledge.
Stanford HAI Healthcare AI Policy Workshop. (2024). *Stanford University: Human-centered Artificial Intelligence. Pathways to governing AI Technologies in Healthcare.*
Steinmüller, W. (1975). Automationsunterstützte Informationssysteme in privaten und öffentlichen Verwaltungen: Bruchstücke einer alternativen Theorie des Datenzeitalters. *Leviathan, 3*(4), 508–543.
Weiser, M. (1993). Ubiquitous computing. *Computer, 26*(10), 71–72.
World Health Organization (WHO). (2023). *Regulatory considerations on Artificial Intelligence (AI) for health*, WHO.
Yuste, R., Genser, J., & Herrmann, S. (2021). It' time for neuro-rights: New human rights for the age of neurotechnology. *Horizons, 18*, 154–164.

Neuroscience, Neuroprotection, and Neurorights: Human Brain Rights and the Brazilian Constitutional Advances

Camila Pintarelli

Abstract Believing it essential to reflect on the expansion of the legal and normative understanding of human dignity in this new digital context, we present this essay to address the necessary recognition that must exist between technosocial dynamism and the human subject, in order to assist in the construction of theoretical bases that demonstrate that scientific and technological development must serve humanity and respect life, equality and freedom. To this end, we will begin our analysis by briefly examining the social evolutionary milestones related to this subject and the ethical impulses derived from them. In the second part, we will examine the current legal stage of the protection of the human mind in the international context. Finally, we will analyze the possibilities that Brazil has regarding the legal treatment of these issues, using a descriptive exposition and attentive to the legislative changes underway in the country.

Keywords Neurotechnology · Human subject · Brazilian constitutional advances

1 Introduction

The fascination with human interaction with robotics has long accompanied society. Simply recalling the drawings and machinery designed by Leonardo da Vinci in the 15th century is enough to confirm that human curiosity has always driven us to seek mechanisms capable of enhancing our capabilities through external instruments or devices integrated in some way with our thinking capacity.

Humanity's fascination with the development of cybernetics has inspired a wealth of literary works, including those by Isaac Asimov, who introduced the famous three laws of robotics. Furthermore, in the legal field, we find innovative works analyzing the subjective entitlement of human rights by robots, as seen in the well-known study

C. Pintarelli (✉)
State of São Paulo's Attorney General's Office, Sao Paulo, Brazil

P. López-Silva (ed.), *Contextualizing Neuroprotection: Latin American Perspectives on the Impact of Neurotechnological Development in Life and Society*, Ethics of Science and Technology Assessment 54, https://doi.org/10.1007/978-3-031-96055-0_10

by Robert Alexy and Alfonso García Figueroa on the famous android Data from the Star Trek series (Alexy & Figueroa, 2007).

In medicine, the prestige of the possibilities of interaction between robotics and human life has been highlighted for many years, from the simple examples of pacemakers and stents to the evolution of neuroscience studies. These began in 1878 with Richard Caton's analysis of an animal brain and quickly advanced to the first human electroencephalogram measurement by Hans Berger in 1929, which is considered a precursor of the so-called brain-computer interface (BCI).

More recently, the emergence of non-invasive neuroimaging techniques has revolutionized the understanding of the human brain (Álamos et al., 2022, 287). This progress has been accelerated with the considerable expansion of the spectrum of available neuroimaging technologies, both in clinical settings and for commercial access. Examples include Magnetic Resonance Imaging (MRI), Computerized Axial Tomography (CAT), Positron Emission Tomography (PET), and Functional Magnetic Resonance Imaging (fMRI).

The rapid development of neurotechnology, with the evolution of connective interface techniques between the brain and machines, is genuinely expanding the boundaries of human action, generating hope and expectations about their positive impacts on the economy and healthcare, particularly in treating and curing diseases. A memorable example is the 2014 FIFA World Cup, where a paraplegic young man took the opening kick-off with the help of the BRA-Santos Dumont 1 exoskeleton, a robotic suit controlled by brain activity through a sensor cap worn on the scalp.

These advances can and should be celebrated as they offer hope for overcoming human fragilities, reaffirming humanity's historical and inherent drive to seek knowledge and improve ourselves. This drive dates back centuries.

However, alongside the undeniable benefits to human life that neurotechnological progress brings, it is also clear that the consolidation of such technologies allows some of their applications to be offered on a commercial scale. This raises ethical and social concerns regarding the consequences of their widespread adoption. In a recent article, professors from Columbia University (USA), the Pontifical Catholic University of Chile, and the Donostia Center for Psychic Research (Spain) expressed concern about the ethical and legal limits involved in the wide range of possibilities for accessing, collecting, sharing, and manipulating brain data (Álamos et al., 2022, 287). This concern has resonated globally.

The advances in neuroscience techniques open an unprecedented possibility for accessing, collecting, sharing, and manipulating information from human brains (…). They can positively impact clinical practice, improving patients' well-being from neurological and psychiatric disorders, offering new preventive, diagnostic, and therapeutic opportunities. However, outside the clinic, Neurotechnology's commercial usages provide new possibilities for self-quantification, cognitive enhancement, personalized communication, and entertainment for regular users. This raises new ethical challenges if misused or inadequately implemented, like the risk of creating distinctive forms of intrusion into people's private lives, potentially causing physical or psychological harm or allowing undue influence on people's behavior without their consent (…) (Álamos et al., 2022, 289).

In this regard, literature already considers the risk that knowledge of neurological structures could become a source of social behavior control—something akin to a true neurocivilization—potentially creating pathways capable of replacing the legal system as a mechanism of social control. Research conducted by the University of Trento, Italy, introduces the concept of "neurolaw," in which legal standards are replaced by neuroscientific standards for social control. These studies suggest the need for constructing international regulatory frameworks to protect humans in such a futuristic scenario (Sommagio, 2022, 287).

Furthermore, neural data analysis methods are becoming increasingly complex due to their integration with artificial intelligence and machine learning tools, enabling algorithms to profile and analyze the information extracted by neurotechnology from the human mind.

A recent study by the Department of Psychology and Behavioral Sciences at Zhejiang University in China analyzed functional magnetic resonance imaging (fMRI) of the brain. Researchers concluded that TikTok's recommendation algorithms can regulate the activity of sub-regions of the brain's "default mode network" (DMN), potentially creating compulsive video-watching behavior (Su et al., 2021).

The facts discussed here highlight ethical concerns and the legal treatment required to preserve the human being in this inevitable context.

Such ethical questions can be initially addressed by introducing philosophical frameworks into the development and enhancement of these research areas, creating ethical standards to guide and illuminate the boundaries related to the interaction of these devices with humans and the use of neural data by other automated decision-making systems.

From another perspective, these same issues can also be addressed through the study and implementation of regulatory policies capable of guiding their use and, more importantly, protecting and preserving the subjective rights involved. This must take into account the interdisciplinary nature of neurotechnology and its natural effects on the legal order, particularly concerning the understanding of human dignity and human rights.

This is precisely the approach we aim to explore in this article. We believe it is essential to reflect on the expansion of the legal and normative understanding of human dignity within this new digital context, to ensure that, when recognizing the existing and inevitable techno-social dynamism with the human subject, scientific and technological development serves humanity and respects life, equality, and freedom.

To this end, we will begin our analysis by briefly examining the social evolutionary milestones related to this subject and the ethical impulses derived from them. In the second part, we will examine the current legal stage of human mind protection in the international context. Finally, we will analyze the possibilities that Brazil has concerning its legal treatment of these issues.

This is a topic of great importance in the legal-academic debate and still carries a considerable degree of novelty, both factually and doctrinally. It reveals not only the necessity of addressing this issue but also the potential for Brazil—as one of the world's leading hubs, markets, and recipients of technological advancements—to assume global leadership by reaffirming the Kantian shift within the context of technological and scientific advancements.

2 Historical Milestones and Regulatory Potentials

As we have seen, human interaction with robotics and technology has been part of social history for a long time. Recent and significant developments in this area, responsible for great revolutions and improvements in favor of human life, are certainly the result of the growing understanding of the human brain and neural activity.

One of the most significant milestones in this regard is the revolutionary BRAIN Initiative—Brain Research through Advancing Innovative Neurotechnologies. On April 2, 2013, under the leadership of then-President Barack Obama, the United States announced the launch of this program, aiming to promote research, development, and application of innovative technologies aimed at creating a dynamic understanding of how the human brain works. The significance of this project is reflected in its team, composed of distinguished experts from leading universities and research centers around the world.

Going beyond current achievements in neurosciences, the BRAIN Initiative specifically aims not only to clarify specific brain areas that underlie the production of conscious mental states but also to describe, with unprecedented precision, the specific neural paths that electrical impulses in the brain take to produce those mental states. This would allow the understanding of how brain processes produce perception, action, memories, thoughts, and consciousness (López-Silva, 2022, 25).

The significance of this program was followed shortly after its launch by concerns about the ethical boundaries of its research and results. President Obama himself expressed these concerns, even sending a query to the Presidential Commission for the Study of Bioethical Issues in the United States, seeking guidance on these issues. He emphasized the possible impacts that discoveries involving the human brain could have on human freedom and equality.

New technologies to better visualize the brain and understand how it works promise to speed the discovery of new ways to treat and prevent brain disorders (…). Advances in neuroscience can also raise ethical and legal issues that require reflection and analysis. (…) We should consider the potential implications of the discoveries that we expect will flow from studies of the brain and some of the questions that may be raised by those findings and their applications—questions, for example, relating to privacy, personal agency, and moral responsibility for one's actions; questions about stigmatization and discrimination based on neurological measures of intelligence or other traits; and questions about the appropriate use of neuroscience in the criminal justice system, among others. It will also be important to consider these types of questions as they relate to different life stages, from infancy through old age (Hagmann & Gerhard, 2014).

This presidential inquiry for interdisciplinary guidelines in the development of BRAIN Initiative research reflects what we previously discussed regarding the potential ethical approaches to the consequences of neurotechnology advancement. That is, it seeks to insert within the empirical procedure ethical boundaries capable of safeguarding human protection.

This goal was underscored in the Commission's response to the inquiry, in which it published the study Gray Matters: Integrative Approaches for Neuroscience Ethics and Society (Hagmann & Gerhard, 2014), offering four recommendations to the White House and the BRAIN Initiative: to explicitly integrate ethics into research as early as possible, to assess existing and future approaches for ethical integration, to incorporate ethics and science into education at all levels, and to include ethical perspectives in review and advisory bodies.

In summary, the Commission concluded that the emphasis in neuroscience should always be on the pursuit of human integration rather than intervention in the human being.

In 2017, countries such as China, South Korea, Japan, Canada, and Australia, along with the European Union, joined the *International Brain Initiative*, with purposes similar to the American BRAIN Initiative, reinforcing the global movement for a qualitative leap in the interaction of technology with humans through knowledge of brain activities. This is driving significant developments in technologies such as Deep Brain Stimulation (DBS), Transcranial Magnetic Stimulation (TMS), and Brain-Computer Interfaces (BCIs), which have shown promising results in the treatment of neurological diseases. In this regard, it is worth highlighting a report made in March 2022 by the prestigious journal *Nature*, about research conducted by the University of Tübingen in Germany and the Wyss Center in Switzerland. A patient in an advanced stage of Amyotrophic Lateral Sclerosis (ALS) received a brain implant that, after one year, allowed the construction of sentences and interaction with the medical team (Chaudhary et al., 2022).

While numerous possibilities for curing and improving human life are being contemplated with neurotechnology advancements, the development and consolidation of the use of these technologies—along with the increasing potential for their commercialization—highlight their societal and individual impacts, which are sometimes difficult to predict.

Over the last years, neurotechnologies such as Deep Brain Stimulation (DBS), Transcranial Magnetic Stimulation (TMS), and Brain-Computer Interfaces (BCIs) are starting to offer promising ways to deal with the burden of specific neurological conditions, such as Parkinson's disease, brain strokes, paralysis, among many others. (…) Worryingly, this neurotechnological revolution seems to be also leading to the creation of commercial and military applications (…). This issue becomes highly problematic in light of the current lack of explicit international regulatory laws for the potential production of such neurotechnological applications (…). This scenario becomes even more complicated if we think that this type of neurotechnologies might allow the development of the unprecedented ability to read minds by decoding, analyzing, and interpreting data about neural activity patterns of human brains, exposing what was once thought to be private, namely our thoughts, beliefs, desires, and (cognitive and behavioral) predispositions. More importantly, the very possibility of recording with such precision the neural activity that produces specific mental states might offer scientists and governments the possibility of not only reading but also controlling the production of mental states in the minds of regular citizens, a process that has been called "brain-hacking" (López-Silva & Valera, 2022a, 2022b, 25).

Beyond the examples mentioned, the commercial use of neurotechnologies brings other practical situations. In the marketing sector, neuroimaging technology is already being used to collect brain data in order to influence consumer decision-making processes, which contributes to the creation of advertising campaigns. In the electoral field, these tools are used to measure the implicit popularity of candidates among voters. This becomes important in light of studies showing that a candidate's initial visual impression has a significant impact, while the development of the campaign and speeches delivered later are of little relevance (Lavazza, 2022, 83).

The situations described indicate that neurotechnology, intensified by its integration with artificial intelligence and machine learning, is daily creating new axiological, anthropological, ethical, and human spaces for action, as it is human self-determination and free will that are at stake.

In this sense, the unprecedented challenges posed by neurotechnology must be interpreted in light of human rights, seeking to understand the effective protection of freedom, equality, and above all, human dignity. This also raises the question of whether the concepts currently present in international regulatory frameworks are capable of accommodating the reality we are experiencing.

Some nuances of scientific and technological development are preserved in certain international documents. One notable example is Article 27 of the Universal Declaration of Human Rights, which guarantees the right of individuals to participate in scientific progress and its benefits. Similarly, Article 15 of the International Covenant on Economic, Social, and Cultural Rights contains a similar provision.

Another important international legal instrument in this context is the Universal Declaration on Bioethics and Human Rights, whose Article 8 refers to respect for human vulnerability and personal integrity. The vulnerability described in this treaty can be interpreted as an inherent human condition, that is, a fragility and corporality that make humans dependent on social relations and the environment for their ontological improvement (Varela 2022, 191–192).

From a philosophical perspective, vulnerability is an inherent condition of human existence in its finitude and fragility, which cannot be overcome or eliminated. By recognizing their vulnerability, individuals understand the vulnerability of others and the need for care, responsibility, and solidarity, rather than the exploitation of this condition by others.

Thus, it is relevant to draw a parallel between the protection provided in the Universal Declaration on Bioethics and Human Rights and the current situation of rapid technological development capable of promoting cognitive enhancements to overcome our individual vulnerabilities. There is a risk of impacting a characteristic that makes us human: human essence in its own vulnerability.

Another important international document is the 1986 Declaration on the Right to Development by the United Nations, which places the individual at the center of economic, social, cultural, and political progress, ensuring their participation and benefit, while promoting and protecting a political, social, and economic order that equally fosters comprehensive human development.

The principles contained in the 1986 Declaration encompass notions of protection that transcend the historical moment of its issuance and reach the technological reality of today, notably equality and inclusion in the context of development. This is particularly important given the potential commercialization of cognitive enhancement tools via neurotechnology. Another crucial point is the transparency and accountability needed to ensure trust in social relationships developed through new technologies.

Despite such references, research indicates that existing international treaties related to this subject do not provide sufficient legal protection to safeguard physical and mental integrity against the development of neurotechnology. This is the conclusion reached by a recent analysis by the NeuroRights Foundation, affiliated with Columbia University in the United States, one of the global pioneers in studies of this nature:

> (…) Existing international human rights treaties are currently unprepared to protect neurorights. Nevertheless, as described in detail in our findings, rapid advances in neurotechnology are no longer science fiction – they are science. It is urgent that the UN play a leading role globally to embrace these exciting innovations while protecting human rights and ensuring the ethical development of neurotechnology (Genser et al., 2022).

This research recognizes that the gaps in international human rights law concerning neurotechnology can be addressed by drawing on expressions or concepts from some existing treaties, as we suggested earlier. For instance, the Universal Declaration on Bioethics and Human Rights and the International Declaration on Human Genetic Data can serve as guidelines for the collection and storage of brain data, as well as indicate baseline standards for human consent in such procedures.

However, the conclusion reached indicates the urgent need for new language and terminologies to be incorporated into international treaties to provide more specific legal protection for the practical situations emerging from the use of various neural technologies, especially those involving the transfer of brain data. This demonstrates that the current concepts of freedom and equality need to be expanded to safeguard human dignity.

The quest for new rights and the enhancement of existing international treaties is a natural and cyclical movement in societies. The recognition that social changes require the formulation of new rights is considered by the best legal scholars as one of the strongest signs of social maturity (Zagrebelsky, 2007, 84).

Bobbio's teachings remain relevant, as he argued that human rights are historical and that their evolution reflects societal needs and technical transformations of humanity (Bobbio, 2004, 18).

In the absence of specific international instruments, global regulatory efforts on the subject, though extremely recent, reinforce the notion of social maturity and the pursuit of new rights. They also indicate the collective awareness of international actors about the need to establish guiding frameworks—binding or otherwise—on the issue.

In this context, the Organization for Economic Co-operation and Development (OECD) can be considered a pioneer among intergovernmental organizations for issuing the *Recommendation on Responsible Innovation in Neurotechnology* in December 2019, the first international regulatory milestone on the subject. This recommendation recognizes the need for international normative standards and was built around nine principles aimed at guiding nations and researchers to anticipate and address the ethical, legal, and social challenges posed by neurotechnology development.

In the same vein, in October 2020, the European Parliament approved a resolution with recommendations on the ethical aspects of artificial intelligence, robotics, and related technologies. In general, it seeks to reinforce social trust and demand transparency and better information flows between citizens and organizations or companies that develop, implement, and use such technologies. This can be considered a major global regulatory leap in the field.

Additionally, some countries have preempted a possible supranational regulatory movement by adapting their domestic legislation to align with new languages and specific human protection needs arising from the current techno-social reality.

An example is Spain, which in July 2021 adopted a *Digital Rights Charter* that includes principles and rights to guide future regulatory public policies to ensure individual and collective protection in new digital environments. Among the points included in the charter are digital rights concerning the use of neurotechnologies, as well as the importance of protecting mental privacy, mental self-determination, and protection against discrimination.

In Latin America, Chile is the only country to date that has advanced in addressing this issue, expressing concern about the need for a regulatory framework capable of recognizing new human rights in light of the technological development currently underway.

By approving Law No. 21.383, published on October 25, 2021, the Chilean Parliament sought to amend the country's Constitution to include the right to neuroprotection as part of the protection of human dignity. This law safeguards the right to mental privacy, identity, and personal autonomy, free will, and self-determination, equitable access to brain cognitive enhancement mechanisms, and protection against algorithmic bias in automated decision-making processes. These protective measures introduced by the Chilean legislature are conceptually classified as *neurorights*.

Specifically, regarding the privacy of information produced by brain activity, the Chilean parliament differentiated it from the already known protection of personal data, clarifying that it is complementary to this other protection. It aims to reinforce fundamental protection of brain information as a true extension of human dignity. Since this information does not strictly constitute personal data, it deserves specific protection, especially to safeguard personal identity and self-determination, which pertain to the core identity of the human being.

These guarantees align with the points made earlier and seek to preserve equality among individuals in a context where intersubjective actions may be driven by technological factors. Above all, they aim to guarantee the freedom of human action.

In addition to these movements, there is the previously mentioned involvement of the UN through UNESCO, which, apart from publishing a report warning about the specific risks of neurotechnologies to human rights, promoted a significant debate in July 2023. The event involved prominent names from science and government worldwide, and the goal was to discuss and evaluate the construction of a global regulatory model for this new reality.

Bringing these concerns and regulatory movements to the Brazilian legal context, it is possible to identify ways to integrate legal protection of the human mind into our legal system. This would strengthen and guarantee other fundamental rights, such as equality and freedom, which require reinterpretation in light of the techno-social context. This may necessitate constitutional updates to align our fundamental law with this new reality.

In this regard, it is worth recalling that the Brazilian Constitution demonstrates the framers' intention to incorporate the concept of human self-determination into the normative framework of human dignity. Indeed, human dignity in the 1988 Constitution is the combination of freedom as a principle of socialization: it affirms the human capacity to govern one's destiny, expressing individuality while simultaneously affirming social rootedness. This is because dignity rests on plurality and collective action (Ferraz Junior, 1997).

This constitutional choice also highlights what was previously mentioned about the presence of vulnerability in the construction of the human being, particularly in the context of human dignity. This shows the importance of the plural interaction of individuals in society to affirm the individual as a holder of their own infinitude.

Our Constitution embraces the idea of human self-determination as a key element of dignity. This is especially relevant in the context of neurotechnology developments and concerns about creating regulatory standards to guide its use in a way that preserves humanity.

In other words, the legal protection of the mind within the Brazilian constitutional framework naturally follows from the conceptual choice of human dignity made by the Constitution. Even though it was drafted at a time when scientific progress to such an extent was not anticipated, the Constitution can accommodate modifications of this magnitude without compromising the fundamental notions that safeguard the rights it enshrines.

A significant step was taken towards incorporating the protection of the human subject within the digital context into the Brazilian Constitution with the Constitutional Amendment No. 17 of 2019. Proposed in the Federal Senate, it added Article 5, Item LXXIX, which provides for the protection of personal data among the fundamental rights and guarantees.

However, we now see that the development of neurotechnologies goes beyond the protection of personal data, reaching the human mind itself, as it is already possible to process and even store neural data. This is not merely about personal data but about mental integrity, which forms the essence of human beings, affecting freedom, equality, personal self-determination, and human dignity.

Considering the conceptual openness conferred to human dignity by the Constitution and the parliamentary recognition of the need for advancements in the provision of fundamental rights and guarantees, as signaled by Constitutional Amendment No. 17 of 2019, we believe it is possible for the list of fundamental rights and guarantees to be further refined. This would reflect Brazil's social maturity by recognizing new axiological spaces for legal protection.

We do not ignore the fact that Brazilian society still lacks the concreteness and effectiveness of numerous fundamental rights already enshrined in our Constitution. The absence of public policies capable of implementing them makes the situation quite concerning in terms of social development and national progress. However, we believe that we cannot cling to this argument to avoid addressing the regulatory necessity of a topic that is increasingly palpable and sensitive in our reality. Its ethical and normative consequences may become apparent only when it is too late to debate them in relation to the protection of the human being.

This is precisely the objective of Proposed Constitutional Amendment No. 29 of 2023, which aims to insert a new item into Article 5 of the Federal Constitution to protect the mental integrity of human beings in the context of scientific and technological advancements. This proposal introduces into the Brazilian constitutional framework the idea of neurorights, already enshrined in Chile. Through this, mental privacy, free will, and self-determination—in other words, mental integrity—would become fundamental rights, making it possible to protect thoughts even before their expression.

It is also worth mentioning that a similar initiative was presented in the state of Rio Grande do Sul, resulting in the Proposed Constitutional Amendment No. 298 of 2023, which aims to elevate the protection of the human mind to the level of public policy guidelines. This proposal was unanimously approved by the Legislative Assembly of Rio Grande do Sul on December 19, 2023.

The inclusion of neurorights as fundamental rights in our Federal Constitution dialogues with the expansion of human dignity in light of technological progress. Beyond that, it allows for the dynamic protection of freedom of thought expression, already enshrined in the same list of rights, by providing specific legal protection to the entire process of something inherent to human essence and social language: thinking and expressing thoughts. Beyond the right to express one's thoughts freely, individuals should be assured the fundamental freedom to think itself.

3 Conclusion

We have seen that the rapid development of neurotechnology, with the evolution of connective interface techniques between the brain and machines, is expanding the boundaries of human action, generating hope and expectation for its positive impacts on the economy and healthcare, especially in treating and curing diseases.

While these advances deserve recognition, it is essential to remain mindful of the ethical boundaries they entail, as this technological leap profoundly affects the mental integrity of human beings. This also encompasses concerns related to the consolidation of automated decision-making processes through algorithmic interactions, which also affect the human mind and demand specific legal protection. Despite the thematic connection, current international legal instruments do not provide sufficient legal protection for the techno-social context we are experiencing. However, there are research initiatives and intergovernmental, supranational, and national efforts—such as those in Chile and Spain—that establish frameworks for protecting mental privacy in the context of the commercial use of neurotechnology. This concern also extends to the Brazilian context. Thus, considering the conceptual framework of human dignity provided by our Federal Constitution, and in light of the paths taken by other global actors, we believe it is possible for Brazilian legislators to sensitize themselves to the need to expand the legal understanding of human dignity in this new digital context. This would involve incorporating into the Federal Constitution similar protections, complementing existing provisions related to personal data, and ensuring that, in the face of the existing techno-social dynamism, scientific and technological development serves humanity while respecting life, equality, and freedom.

References

Álamos, M. F., Kausel, L., Baselga-Garriga C., Ramos P., Aboitiz F., Uribe-Etxebarria X., Yuste, R. A Technocratic Oath. In P. López-Silva, & L. Valera (Eds.), *Protecting the mind: Challenges in law, neuroprotection, and neurorights* (pp. 99–113). Springer

Alexy, R., & Figueroa A. G. (2007). Star Trek y los derechos humanos. Tirant lo Blanch.

Bobbio, N. (2004). *A Era dos Direitos*. Elsevier.

Chaudhary, U., Vlachos, I., Zimmermann, J. B., Espinosa, A., Tonin, A., Jaramillo-Gonzalez, A., Khalili-Ardali, M., Topka, H., Lehmberg, J., Friehs, G. M., Woodtli, A., Donoghue, J. P., & Birbaumer, N. (2022). Spelling interface using intracortical signals in a completely locked-in patient enabled via auditory neurofeedback training. *Nature Communications, 13*(1), 1236. https://doi.org/10.1038/s41467-022-28859-8

Ferraz Junior, T. S. (1997). Direito e Cidadania na Constituição Federal. Revista da Procuradoria Geral do Estado de São Paulo. São Paulo, n. 47/48.

Genser, J., Herrmann, S., & Yuste, R. (2022). *International Human Rights Protection Gaps in the Age of Neurotechnology*. Disponível em: https://static1.squarespace.com/static/60e5c0c4c4f37276f4d458cf/t/6275130256dd5e2e11d4bd1b/1651839747023/Neurorights+Foundation+PUBLIC+Analysis+5.6.22.pdf

Hagmann, P., & Gerhard, S. (2014). *GRAY MATTERS: Integrative approaches for neuroscience, ethics, and society* (Vol. 1). Washington.

Lavazza, A. (2022). Free will and autonomy in the age of neurotechnologies. In P. López-Silva, & L. Valera (Eds.), *Protecting the mind: Challenges in law, neuroprotection, and neurorights* (pp. 41–59). Springer.

López-Silva, P., & Valera, L. (2022a). *Protecting the mind: Challenges in law, neuroprotection, and neurorights*. Springer.

López-Silva, P., & Valera, L. (2022). The concept of mind in neuroprotection debate. In P. López-Silva, & L. Valera (Eds.), *Protecting the mind: Challenges in law, neuroprotection, and neurorights* (pp. 9–19). Springer.

Sommagio, P. (2022). Neuroscience, neurolaw and neurorights. In P. López-Silva, & L. Valera (Eds.), *Protecting the mind: Challenges in law, neuroprotection, and neurorights* (pp. 71–87). Springer.

Su, C., Zhou, H., Gong, L., Teng, B., Geng, F., & Hu, Y. (2021). Viewing personalized video clips recommended by TikTok activates default mode network and ventral tegmental area. *NeuroImage, 237*, 118–136. https://doi.org/10.1016/j.neuroimage.2021.118136

Valera, L. (2022). Mental integrity, vulnerability and brain manipulations: A bioethical perspective. In P. López-Silva, & L. Valera (Eds.), *Protecting the mind: Challenges in law, neuroprotection, and neurorights* (pp. 99–113). Springer.

Zagrebelsky, G. (2007). El derecho dúctil. Trotta

Neural Data and Neuroprotection in the Chilean Constitution Reform Bill: Critical Considerations

Pablo López-Silva and Raúl Madrid

Abstract In October of 2021, Chile became the first country in the world to approve a constitutional reform that explicitly integrates legal safeguards against the misuses of medical and commercial neurotechnological applications (Law N° 21.383). This legislative initiative was aimed to show the Chilean government's explicit concern about how neurotechnological progress could negatively impact society if not properly regulated. Due to its novelty and nature, this constitutional amendment has been the target of heated debate over the last years. The aim of this chapter is twofold. First, after specifying the ethical concerns that gave rise to the Chilean constitutional reform, we clarify the conceptual and legal foundations contained in the core message of the bill proposed and discussed by the Senate of Chile (Bulletin 13.827–19). After, we offer some critical considerations about the content of the bill and the way in which it is aimed to incorporate the so-called 'neurorights' to the Chilean legal order.

Keywords Neuroprotection · Neurorights · Chilean Constitution

P. López-Silva (✉)
Escuela de Psicología, Universidad de Valparaíso, Valparaíso, Chile
e-mail: pablo.lopez.silva@gmail.com

Instituto de Sistemas Complejos de Valparaíso, Valparaíso, Chile

Centro de Investigación del Desarrollo en Cognición y Lenguaje, Universidad de Valparaíso, Valparaíso, Chile

The Collaborating Centre for Values-Based Practice in Health and Social Care, St Catherine's College, Universidad de Oxford, Oxford, UK

R. Madrid
Facultad de Derecho. Facultad de Filosofía, Law, Science & Tecnology Program, Pontificia Universidad Católica de Chile, Santiago, Chile

P. López-Silva (ed.), *Contextualizing Neuroprotection: Latin American Perspectives on the Impact of Neurotechnological Development in Life and Society*, Ethics of Science and Technology Assessment 54, https://doi.org/10.1007/978-3-031-96055-0_11

1 Neurorights and Threats of Neurotechnologies

Collaborative research projects in medical neurosciences have led to the development of novel methods for recording and manipulating neural activity in human brains with unprecedented specificity (Alivisatos et al., 2013; Koroshetz et al., 2018; Ngai, 2024).[1] This progress has allowed the development of new treatments for neurological conditions such as Parkinson's disease, strokes, and brain paralysis using neurotechnological technics such as Deep Brain Stimulation (DBS), Transcranial Magnetic Stimulation (TMS), and Brain-Computer Interfaces (BCIs). Neurotechnological applications in medical science initially motivated a number of ethical concerns about how neural data should be recorded, managed, stored, and used in therapeutic contexts. However, technological progress in medical science paved the way for the creation of commercial neurotechnological applications in the private sector, and, with this, ethical worries about the legal status of neural data extended beyond the medical domain.

In 2017, the *MorningSide Group*—a group of researchers associated with the US Brain Initiative—expressed some worries about how medical and commercial neurotechnological applications could allow a type of *mindreading* by decoding, analyzing, and interpreting data about neural activity patterns of human brains. Rafael Yuste, leader of the *MorningSide Group,* claimed that neurotechnologies with this type of specificity could expose the private sphere of thoughts, beliefs, desires, and (cognitive and behavioral) predispositions. What was always thought to be private and inaccessible, now could become exposed (López-Silva, 2022). The scenario offered by neurotechnological progress could exacerbate social inequalities and "offer corporations, hackers, governments or anyone else new ways to exploit and manipulate people" (Yuste et al., 2017, 160). One of the most quasi-science fictional worries in this context is that the ability to record and reproduce specific brain patterns responsible, for example, specific bodily movements could allow actual real control of people through the re-activation (*via* cortical stimulation) of those patterns in the subject's brain by using, for example, a smartphone or any other device able to produce controlled and fine-tuned electric discharges. Yuste et al. (2017) call this possibility, 'brain-hacking'.

Considering these ethical worries, the *MorningSide Group* pioneered the idea that neurotechnological—medical and commercial—progress should be accompanied by the development of legal safeguards aimed at protecting "people's privacy, identity, agency, and equality" (Yuste et al., 2017, 159). The guiding idea here was that current legal frameworks were not equipped to deal with the specific and new type of threats posed by neurotechnological to the privacy of neural data. These safeguards have been

[1] In 2013 the U.S. BRAIN Initiative is stablished in order to "accelerate the development and application of new technologies that will enable researchers to produce dynamic pictures of the brain that show how individual brain cells and complex neural circuits interact at the speed of thought" (Koroshetz et al., 2018, 1). In 2017, similar enterprises from China, Korea, the European Union, Japan, Canada, and Australia joined the so-called *International Brain Initiative* aiming at cooperatively exploring new treatments for neurological and psychiatric disorders (Adams et al., 2020).

generically called *neurorights* in the literature and they seem to express "intentions to protect people from the eventual dangers of neurotechnology" (Zúñiga-Fajuri et al., 2021, 166). A number of specific neurorights—*qua* safeguards—have been proposed in the literature. Initially, Ienca and Andorno (2017) introduced the notion of neuroright in their analysis of human rights in the age of neurosciences. The authors will claim that "the possibilities opened up by neurotechnological developments and their application to various aspects of human life will force a reconceptualization of certain human rights, or even the creation of new rights to protect people from potential harm" (2017, 2). Ienca and Andorno (2017) operationalize their worries by defending Bublitz's right to cognitive liberty i.e. "the right to alter one's mental states with the help of neurotools as well as to refuse to do so" (Bublitz, 2013, 234). Arguing that currently there are no legal protections from having your mind involuntarily read, Ienca and Andorno (2017) also proposed the creation of a 'right to privacy' under Wolpe's (2009) idea that:

> The skull should be designated as a domain of absolute privacy. No one should be able to probe an individual's mind against their will. We should not permit it with a court order. We should not permit it for military or national security. We should forgo the use of the technology under coercive circumstances even though using it may serve the public good (Page).

Ienca and Andorno's (2017) original proposal also includes the creation of a right to mental integrity indicating that current laws of protection of integrity makes no explicit reference to neurotechnological practices. In the same vein, the authors propose the creation of a right to psychological continuity based on the idea that changes in brain function caused by brain stimulation could cause alterations in mental states critical to personality and could thereby affect an individual's personal identity (see also, Decker & Fleischer, 2008). There are a number of other proposals in current literature—for example, Ratoff (2024) has recently proposed the creation of a right to mental autonomy—however, a systematic examination of all of them would take us too far from the rather humble aim of this chapter.

What is important here is that even though the concept of neuroright has become a popular topic in current neuroethics, the notion has been heavily disputed. For several authors, the notion of neuroright is replete with philosophical misunderstandings and unrealistic—if not mistaken—assumptions about the real power and possibilities offered by current neurotechnology (Bublitz, 2022; Gilbert & Russo, 2024; Gilead, 2015; López-Silva & Valera 2022; Susser & Cabrera, 2024; Zúñiga-Fajuri et al., 2021). For example, Zúñiga-Fajuri et al. (2021) suggest that discussing properties such as liberty, privacy and liberty when it comes to brains and neural data is simply a case of a *mereological fallacy,* namely, the tendency to attribute to brain properties that only belong to persons. They add that neurorights are a case of normative redundancy since they aim at protecting old rights from new threats. The idea is that in the same way in which new forms of endangering someone's life would not require new versions of the right to life, new threats to freedom, privacy, and so on would not require new versions of these rights. Here we shall not examine the plausibility of these claims further. However, we think that there are epistemic and scientific reasons to take the threats posed by neurotechnological applications very seriously (López Silva & Madrid, 2022; López-Silva et al., 2024). Instead, in the remainder of this chapter we discuss the first attempt in the world at establishing and integrating neurorights into an existing legal system.

In October of 2021, Chile became the first country in the world to approve a constitutional amendment that explicitly protects citizens against the misuses of medical and commercial neurotechnologies (Law No. 21.383). Due to the influence of the *MorningSide Group* in the Chilean political landscape, this reform was intended to show the Chilean government's explicit concern about how neurotechnological progress could negatively impact society if not properly regulated. Instead of discussing the general problems associated with the notion of neuroright our focus here is the examination of the conceptual and legal foundations contained in the core message of the pioneering constitutional reform bill aiming at protecting Chilean citizens from neurotechnological threats (Bulletin 13.827–19).[2] After this, we offer a critical examination of the content of the bill and the way in which the constitutional amendment is aimed to incorporate the so-called 'neurorights' to the Chilean legal order.

2 The Neuroprotective Constitutional Reform and Its Core Message

The Chilean constitutional reform bill was proposed by senators from different political parties on October 7, 2020 (Bulletin No. 13.827-19). One of the ideas that motivated this action was that new developments in the neurotechnological field could entail a greater capacity for individuals to control different aspects of the brain and behavior. Therefore, neurotechnological progress could potentially become a threat to, among others, a subject's *autonomy* and *freedom*. An underlying idea in the reform bill is that greater understanding of how the brain works should lead to discussion about the kind of surveillance that a subject's neuronal activity should be allowed. The argument in the message of the bill is similar to that made by Persson and Savulescu (2008) to justify moral enhancement, but, in this context, applied to neuroprotection: since the decision to understand how the brain works has already been made—and may even benefit individuals with a number of neurological conditions—it is necessary to prevent abuses (Senate of Chile, Bulletin No. 13.827-19, 2).

One justification for the constitutional reform proposal is that none of the current normative frameworks on human rights specifically refer to the protection of the autonomy and psychic identity of persons, making it necessary to do so in a positive manner. The same need justifies the MorningSide Group's proposal to incorporate amendments in legal frameworks that specifically protect neurorights,

[2] The full bill can be accessed here: www.senado.cl/appsenado/templates/tramitacion/index.php?boletin_ini¼13827-19 [last visited 12 September 2024].

through high-level international instruments that define criminal offenses related to the neurotechnological and artificial intelligence fields in a manner similar to the prohibitions listed in the *International Convention for the Protection of All Persons from Enforced Disappearance* (2010). It is important to point out that the Article 11 of the *American Convention on Human Rights* has expressly recognized issues key for the debate on Neurorights: *1. Everyone has the right to have his honor respected and his dignity recognized. 2. No one may be the object of arbitrary or abusive interference with his private life, his family, his home, or his correspondence, or of unlawful attacks on his honor or reputation. 3. Everyone has the right to the protection of the law against such interference or attacks.* Based on these examples, the *Inter-American Court of Human Rights* has also concluded that there is a very close relationship between dignity and the right to physical and mental integrity, understanding the integrity of the human being as a concrete manifestation of that dignity. According to the authors of the message contained in the reform bill, this point would be fundamental in explaining the placement of the right to neuroprotection at the very core of constitutional guarantees.

A second important situation for the introduction of neurorights within the Constitution—according to the bill—is the role of new neurotechnologies in the contemporary world. This issue is of utmost relevance as technological progress seems to stand as a paradigm of meaning in contemporary culture. However, the use of technology is not new for humans, it has been present since our very beginnings. The 19th century, like the 21st century, also proclaimed technology as a key explanatory factor for its social and cultural progress, and surely the same has happened at other times in history. What, then, makes contemporary neurotechnological progress so exceptional compared to previous technological progress? The answer seems to lie in the ability to record and manipulate a certain specific type of information, as well as in the nature of the type of data that today can potentially become public, namely, neural data. As a result of this, some boundaries that previously seemed unquestionable and firmly established, such as that of the human mind, now appear "ambiguous" (Madrid et al., 2021). Challenging natural limits always carries the risk of manipulating aspects of reality that are not truly willing to submit to our control, as in *Goethe's The Sorcerer's Apprentice*. For this reason, the reform bill message proposes that:

> The constitutional enshrinement of the right to neuroprotection stems from the need to protect human dignity against the use of new techniques, especially concerning the protection of the 'human brain,' a concept that is not limited to a mere physical dimension, but rather expands towards its dimension of mental potentiality, which encompasses the mysteries of human existence, and for that reason, it must receive the highest fundamental legal protection (8).

Two challenges arise here. First, the need for clarification of what the right to neuroprotection consists of (positive challenge) and second, to understand the limits of this right and how it specifies the actions of third parties (negative challenge). As Nogueira (2005) reminds us, the limit of a right is the boundary between what something is and what it is not; it is part of the structure of the right and considers all other constitutionally protected rights and interests. The boundary of a right presupposes, then, the existence of a constitutionally protected content, predefined, within which it implies a contour. In this regard, the reform bill argues that the content of the right to the protection of neural data is "multiple", meaning that it includes a set of prerogatives that, on the one hand, translate into the power to reject arbitrary attacks or transgressions, and, on the other, require positive actions by civil authorities to ensure its protection. This argument is based on the notion of "constitutive function" of a law proposed by Zagrebelsky (2018) who states that:

> The claim of rights is a manifestation of the strength and maturity of a society, a sign of progress toward a goal represented by human autonomy in the face of any social reality that seeks to be accepted as 'given' (82).

It is important to point out that the existence of natural rights prior to the State or the Constitution does not contradict the dialectical nature of human knowledge but rather confines it to the realm of things that, by their very nature, can be modified by the hermeneutic perspective of legal discourse. This would exclude matters that—by their very essence—lie beyond human will, which would be precisely those universal legal demands applicable to all humans, derived from their essence. About this, the current Chilean Constitution explicitly states that: "The exercise of sovereignty is constrained by the respect for the essential rights that emanate from human nature" (Article 5, Section 2).

3 Neurorights in the Chilean Constitutional Reform Bill

Following Yuste's MorningSide Group proposal, the Chilean Constitutional reform bill identifies four ethical elements that constitute the essential content of neurorights. This proposal cannot be considered exhaustive as neurotechnological developments in the future may add aspects that cannot be currently configured either theoretically or normatively. In other words, the proposal is that the core of the fundamental neuroprotective guarantee would consist, at a minimum, of the following four aspects:

(a) The right to privacy of the information produced by brain activity which can be accessed through neurotechnology (*Privacy and Consent*): this refers to the protection of neurodata, which the bill characterizes as "useful and valuable information". The argument is that, without proper safeguards and security measures, this information could pave the way for the annulment of privacy or informational self-determination. It is noted that in Chile, through Law No. 21.096, the Constitution incorporated the right to personal data protection as a fundamental guarantee. The authors of the reform bill believe that the proposed modification would strengthen the protection of information about the brain; this is the purpose of the reform: an explicit legal safeguard that makes an existing protection more powerful and effective. This is deemed necessary in order to protect human dignity, as it is not merely a matter of personal data but rather a category of information so personal and fundamental that it must be explicitly protected.

(b) The right to personal identity and self-determination (*agency*): neurotechnology indeed opens the possibility of nullifying or altering a person's identity, despite the inherent ambiguity of the concept. While it allows, or may allow, the treatment of previously incurable diseases, it also poses a risk to personal identity, as it could become a tool for inhibiting consciousness and the determination of the "self" (through the aforementioned "mental hacking"). In other words, neurotechnology could diminish consciousness or create amnesic gaps, among other unintended effects. The issue of mental identity is probably the most theoretically important from the perspective of the human person faced with detailed knowledge of the brain and nervous system, touching on debates such as the spiritual versus purely material nature of individuals. The concept of identity used in the reform bill will be closely related to the foundation provided for the higher activity of the mind. Thus, one's self-understanding will differ depending on whether thought, feelings, etc., are considered merely the result of brain and nervous system activity or stem from a non-material/spiritual principle.

(c) The right to equality in the face of cognitive enhancements (*cognitive augmentation or enhancement*). The reform bill emphasizes the need to regulate the use of new neurotechnologies to avoid potential attacks on justice. Among the range of enhancements that can be applied to humans—i.e., alongside merely therapeutic interventions—are neuroenhancements, which correspond to the expansion of cognitive, emotional, and motivational functions in healthy individuals through drugs or other means. Strictly speaking, cognitive enhancement is not therapeutic: it does not involve reversing or compensating for deficiencies in intellectual function caused by mental illnesses, neurological disorders, or their decline in old age. The situation is different if the goal is to increase performance beyond its "normal" state (López-Silva, 2021).

(d) The right to control algorithmic biases (*bias*). Due to technological advancement, the authors of the bill argue that many everyday decisions are made by machines through artificial intelligence (AI) algorithms. These processes include personnel selection, contract signing, acceptance of terms, information searches, money transfers, and many other operations organized and articulated through algorithms. Bias occurs when the model makes inappropriate correlations, such as forming false connections between data nodes. It can also involve selection bias, which occurs when the database is not representative of the diversity present in the social environment. The Project attributes an ideological content to bias; however, by assuming that "*some of these algorithms discriminate against minorities,*" they sometimes operate by amplifying the trends reflected in the databases. Since many technologies applied to understanding the human mind are based on algorithms, and the biases in their design and application would be directly executed in the brain, it is essential to establish the necessary legal safeguards against bias in automated decision-making. For this reason, they argue, the bill that regulates the protection of personal data (Combined Bills No. 11,144-07 and No. 11,092-07) seeks to enshrine the right to object to automated personal assessments and the principle of privacy by design (Ph.D.) as mechanisms to address potential biases in the development and design of algorithms. However, the "problem of bias" is, in reality, the problem of "prejudice" (i.e., the existence of a prior judgment that affects or modifies the subsequent one). This points to a prior debate: whether the knowledge possessed by human beings originates from evidence or is the result of a method, that is, a process of demonstration. The classical answer to this problem was that the starting point of knowledge was external reality. In this context, prejudice—meaning the judgment prior to formal thought—was not only not considered a defect but pointed to the very condition of human cognitive structure.

4 The Chilean Constitutional Reform Bill: Critical Considerations

The proposed constitutional modification is a single article that involves the incorporation of a new section into Article 19, No. 1, of the Constitution (physical and mental integrity), although it is also related to a number of other already existing fundamental guarantees. The aim of the bill adding "some essential elements for the proper protection of human rights in the face of the development of neurotechnology." On September 6, 2021, a discussion committee was established by the Honorable National Congress, initiated by the motion of Senators Girardi, Goic, Chahuán, Coloma, and De Urresti. This text is included in Law No. 21.383 from the Ministry of Science, Technology, Knowledge, and Innovation, which was published in the Official Newsletter on October 14, 2021. The text proposed reads as follows:

> Law No. 21.383: Amends the fundamental charter to establish scientific and technological development in the service of people.

Unique Article: The number 1 of Article 19 of the Political Constitution of the Republic is amended as follows:

1. In the current final paragraph, the semicolon is replaced with a period.
2. The following new final paragraph is added: 'Scientific and technological development shall be in the service of people and shall be carried out with respect for life and physical and mental integrity. The law will regulate the requirements, conditions, and restrictions for its use on individuals, particularly safeguarding brain activity, as well as the information derived from it.

4.1 On the Need and Justification for a Constitutional Reform

There are two general issues here. The first has to do with whether it was technically preferable to incorporate the neurorights into number 1 of Article 19 of the Constitution, or to have made it a specific section within the general catalog of rights that the article enshrines (López-Silva & Madrid, 2021). It seems evident that the decision was made to associate these new prerogatives with the idea of "psychic integrity" contained in number 1. Nevertheless, this is a methodologically debatable issue. Its presence in number 1 reinforces the value of personal unity and unicity even in the realm of consciousness, which, in our opinion, is a positive effect. The second general issue has to do with whether the reform was really necessary. Fundamental guarantees are, in themselves, legal principles and, therefore, involve a core set of non-explicit norms. In other words, the development of what is implicit in one of these guarantee-principles does not affect its material content, and consequently, everything contained in its core is also protected. Article 19, No. 26, of the Chilean Constitution guarantees everyone:

> The assurance that the legal provisions mandated by the Constitution that regulate or complement the guarantees it establishes, or that limit them in cases where this is authorized, will not affect the rights in their essence, nor impose conditions, taxes, or requirements that impede their free exercise.

This is also linked to Article 5, inc. 2, which states that:

> Sovereignty is limited by the essential rights that emanate from human nature [thereby establishing that] the limit on the content of rights extends not only to the legislator but also to the instituted constituent power, establishing a guarantee of irreversibility in the field of fundamental rights.

The essential content of privacy protection is enshrined in Article 19, No. 4, which states:

> The Constitution ensures respect for and protection of the private life and honor of individuals and their families, as well as the protection of personal data. The processing and protection of this data shall be carried out in the manner and under the conditions determined by law.

This text incorporates the modification of Law No. 21,096 of June 16, 2018, which explicitly granted constitutional status to the protection of personal data in Chile. Prior to this law, the protection of such data was considered included in Article 19, No. 4, of the Constitution (private life). This creates a special legal reserve (personal data) in relation to the general one established in Article 19, No. 26, which makes its processing and protection a matter of law. Personal data had previously been defined in Law No. 19,628, Article 2, as "*any information concerning identified or identifiable natural persons.*

Thus, the guarantee explicitly incorporated into the Constitution by Law No. 21,096 seems to provide sufficient protection for the type of private information in question, without the need to add anything. For this reason, the proposed constitutional amendment in the project bill we are discussing lacks a clear substantive technical-legal justification. It should also be noted that constitutional guarantees include a clause known as the "eternity clause" (*Ewigkeitsklausel*), meaning that they cannot be modified in spirit even by the agreement of all members of the Chilean Congress (parliament). From a hermeneutic perspective, fundamental guarantees are also an open catalog: "[…] by their very nature, they consider the possibility of new attributions and duties arising from the major focal points of meaning that constitute their various provisions. Open catalogs are enriched through the interpretation of existing norms". In this way, one could say that Neurorights—at least in Chile—are subsumable within the existing fundamental guarantees and do not require a constitutional amendment like the one proposed. Now, we will proceed to explore more specific issues regarding the unique article of the proposed constitutional reform in question.

4.2 Critical Comments on the Core Message of the Constitutional Reform Bill

As seen, the final version of the constitutional reform did not include terms like "mental or neural integrity" or "brain indemnity", which were present in the original draft (Bulletin No. 1328-19). The final decision, which we believe is correct, recognized that these concepts are part of an ongoing scientific and epistemological debate (Borbón & Borbón, 2021; López-Silva 2022a). The complexity of these issues, particularly regarding whether human consciousness is purely neurobiological or also includes a nonmaterial dimension responsible for higher cognitive activity cannot yet be adequately translated into legal norms. Acknowledging the significance of these discussions, the decision during the bill project's legislative process was to adopt a "more pragmatic and broad" approach. This pragmatic solution avoids locking the legal framework into unresolved doctrinal debates, instead opting for a flexible approach that can evolve as scientific understanding progresses.

The specific wording of the bill directly emphasizes the State's duty to safeguard "brain activity" within the general declaration that biotechnological improvements must be functional for individuals and respect their fundamental rights, such as life and integrity. This approach aimed to avoid less specific proposals that focus on protection in the terms previously established—specifically without explicitly including technologies that enable the recording of neuronal information or the potential unlawful manipulation of personal decisions. Furthermore, the expression "restrictions" was added to the matters delegated to the law, granting the legal norm the prudent authority to limit the scope of interventions covered by the bill project. The report from the Joint Commission that examined the bill noted that normative conditions can generally be supplemented through regulations because these regulations have sufficient "normative density" (referring to the ability to legally protect both those who develop the technologies and those who use them or suffer their effects). Here the intention was for the modifications to maintain the neutrality of technologies, based on the assumption that the tools developed by humans are not inherently good or bad but only become so based on their instrumental use. However, the law establishes prohibitions or restrictions on certain uses that require special safeguards, rather than on neurotechnologies themselves. This distinction ensures that while technologies can be beneficial, there is still oversight and regulation to prevent misuse that could infringe on individual rights.

In the third stage of discussion of the bill, the Chamber of Representatives removed the term "inclusive" from the final text, which referred to the incorporation of brain activity and the information derived from it. The rationale provided for this change was that the term seemed to separate brain activity—and the information obtained from it—from other activities that can be conducted regarding physical and mental integrity. By eliminating the term "inclusive," the lawmakers aimed to prevent the formation of an interpretive criterion that would exclude other matters. The reasoning was that if these elements required a special clause to be included within physical and mental integrity, it might imply that other activities not explicitly mentioned would be excluded from protection. In place of "inclusive," the word "especially" was introduced, conveying that neural activity represents a specific mode of protection for integrity, rather than being a separate "genre" included externally. This adjustment was intended to ensure that all aspects of integrity—physical and mental—are protected under the same umbrella, reinforcing the comprehensive nature of the safeguards established in the constitutional reform.

Some earlier versions of the bill explicitly mentioned the desire to eliminate or mitigate the "anthropocentric" nature of Article 19 of the Constitution. The idea was to highlight that the development of science and technology is not solely rooted in human well-being, but can also occur independently of it, as seen in the case of environmental protection. This proposal suggested that the environment would hold intrinsic value, considered abstractly, rather than merely in terms of its functionality for human development. This thesis aligns with a view of humanity—as a legal subject—that cannot be qualitatively distinguished from things or goods, thereby calling into question the idea that the state must serve the people established emphatically in Article 1, paragraph 4, of the Constitution, which states that, "The State is at the service of the human person, and its purpose is to promote the common good". As Soto Kloss (2017) argues, this principle represents a "legal duty" imposed on the State by virtue of its purpose and the accidental and instrumental character it possesses; it serves as a means of perfecting individuals. Thus, the Constitution implies an instrumental conception of the State, which is not seen as an end in itself but as a means or instrument whose aim is to foster the integral development of human beings. This perspective encourages a broader interpretation of the role of the State in relation to the environment and other non-human entities, emphasizing the need for a balance between human interests and the intrinsic value of the natural world. The entire section on fundamental guarantees in the Constitution is currently protected by the principle of servility. Therefore, the declared intention to avoid "anthropocentrism" is odd, given that the proposed norm specifically seeks to protect the various attributions of individuals as legal subjects recognized by the Fundamental Charter and the existing legal framework.

The bill's focus on individual rights aligns with the Constitution's overarching goal of promoting human dignity and well-being, suggesting that any move away from anthropocentrism must still grapple with the foundational purpose of the legal system, which is to serve and protect the interests of human beings. In this sense, the proposed reforms, even when considering broader ethical implications, ultimately return to the centrality of the individual within the legal structure. Thus, while there may be a desire to expand the scope of protection beyond human interests to include environmental and non-human considerations, the challenge remains in balancing these interests with the established legal framework that prioritizes human rights and dignity. This ongoing debate highlights the complexity of redefining the role of the State and the legal system in a way that respects both human and non-human rights while adhering to the foundational principles of the Constitution.

The bill project constitutes a form of recognition of human nature and its integrity from a neurological perspective, thereby explicitly adding a new dimension to these notions, resulting from technological advances that allow for a better understanding of brain function. Thus, the constitutional reform in question can be understood as an explicit articulation of the essential content of the guarantee, representing a development related to the space-time determinations for the realization of rights, in light of the new technical possibilities offered by scientific and technological advancements. The final text does not identify the specific rights associated with the protected good, as is the case with other guarantees contained in Article 19 of the Constitution. It is reasonable to assume that the aforementioned classification developed by Yuste and his team from the Morningside Group has been kept in mind. From a theoretical standpoint, the problem remains that these notions are still in a process of development and configuration, such that there is no consensus regarding their content, limits, and characteristics (Bublitz, 2022). Therefore, it is complex to determine, at this stage of neurotechnological development, for example, what exactly concepts such as "neural identity" mean, nor what their potential legal effects might be. This issue appears to be one of the most significant objections when drafting normative texts that do not contain overly open or diffuse elements, and it is deliberately overlooked to avoid theoretical discussion, as mentioned earlier.

The text of the reform bill leaves room for the jurisprudential development of the concept of neurorights, notwithstanding all that has been explained in its core message. If the issue is not resolved at the legislative level, it would likely be transferred to the Courts of Justice. The main problem is that within the same doctrine, there is no general agreement on what exactly constitutes a neuroright, nor is there complete precision regarding a number of other notions used, such as, for example "cognitive enhancement" (and many others). Here, it is also worth questioning whether this newly enshrined right is an autonomous right or whether it is linked to other related rights. The protection of 'mental' or 'neural' privacy, in principle, can be conceived as being connected to the right to life and physical and mental integrity, as determined by the Project in question (art. 19, No. 1). However, it is evident that it also has a specific connection to the protection of personal data, as noted above (art. 19, No. 4). Furthermore, we believe that the right to one's own image should also be included here, a matter that is implicitly protected in the Chilean Constitution, either because it is an aspect of the external dimension of the right to respect for private life ("*constituting the most external aspect, which also guarantees an area of autonomy and control over one's most characteristic and defining attributes, an irreducible and inherent possession of the person*") or because it is an essential right of the individual, possessing an autonomous character, even if it is related to privacy in a broad sense. Additionally, a relationship with equality before the law (art. 19,

No. 2) can be observed, due to the countless issues that could arise from a partial or inequitable process of neurotechnological intervention. Finally, there is also an evident connection to personal autonomy or freedom (art. 19, No. 7).

4.3 Consent and the Constitutional Reform Bill

Some issues arise also regarding the type of consent required for the use of neurotechnologies. The original modification included consent as a prerequisite for the use of neurotechnological interventions. Boxing (sport) was cited as an example. In boxing individuals consent to participate in an activity in which they could potentially be injured, as this would be analogous to the situation at hand:

> Individuals can perfectly consent to engage in contact sports such as boxing [...]. These generate profound interventions in the protective sphere of Article 19, No. 1, and are considered fully lawful when conducted within the framework of what is socially acceptable and do not violate their essential content. The same provision should apply to the impact on neuro-rights.

We share this opinion, but for the purposes of evaluating the nature of consent in these matters, the views of authors such as Agar (2014) are relevant. Agar (2014) argues that experiences can only be understood within the meaningful horizon of the normal range of human experience; that is, it is not possible to connect with experiences that are too far removed from the margin that constitutes genuine human experience. Therefore, consent for certain types of neural enhancements could be problematic because consent cannot truly be given regarding what is unknown, this, because it lies beyond the capacity for understanding in a hermeneutic sense and would thus constitute invalid consent. It is possible that such interventions may not require a prohibition—this would need to be discussed—and that reckless acts could be permitted as long as they do not harm others, with consent in these matters being akin to excessive alcohol consumption. Thus, enhancements that can be predicted to be detrimental to the individual could be allowed. This remains a hypothetical open problem that will need to be analyzed in due time. Nevertheless, there is a significant difference between "allowing" biotechnological human enhancements and establishing them as a "right." The latter option entails several theoretical and practical issues, including the potential obligation of the state to guarantee and provide such enhancements for citizens who request them.

In subsequent discussions, the reference to consent was removed from the text of the reform, understanding that the Constitution currently does not address or refer to this concept, which is treated at the level of the Civil Code carrying legal rank. Moreover, it was considered that the neurorights protection project already refers to three distinct and specific forms of consent: a generic form, another applicable to the medical or therapeutic field, and finally, one applicable to the realm of research. On another note, in the Report of the Joint Commission of Representatives, representative Tohá indicated that, unlike other laws or reforms, this project represents a development or progress that is not finite, whose limits and consequences are unknown. Therefore, the due consent of the affected individuals constitutes a fundamental safeguard that the project should include. The advisors of Senator Coloma recommended that consent should be regulated merely at the legal level rather than constitutionally. Currently, they argued, the law establishes specific consent systems for certain procedures (such as those related to the Scientific Research Law or the Law on Patients' Rights and Duties), which contain exceptions or safeguards for emergency situations in which a person cannot give their consent, a matter that should be respected by not including the issue in the constitutional statute. Additionally, with the referral to the law, consent would already be regulated. Furthermore, it is noted that the Bill for the Protection of Neuro-Rights specifically regulates consent, providing a general rule applicable to any type of neurological intervention, including cases of scientific research and medical treatment. Carlos Amunátegui indicated that not constitutionalizing consent allows for the protection of aspects of personal autonomy and freedom of conscience. Lorena Donoso emphasized that the discussion pertains to the realm of fundamental guarantees and, therefore, to the inalienable rights that stem from human dignity and the essence of the person. She added that it would be inappropriate to consider rules of consent within the framework of these constitutional guarantees, as protecting individuals' lives cannot wait for their consent, and there would be cases where mere consent would not be sufficient. The law would thus be tasked with establishing the conditions and requirements for the application of neurotechnologies, among which is consent.

There is, therefore, a delegation to the law to limit the right and to specify in the future which rights are concretely associated with neuroprotection. Given the nature of the rights in question, it may be advisable to require an additional quorum for this law. However, there is another issue related to consent, but in the legal realm. The neurorights law currently under consideration prohibits, in its Article 3, certain interventions on patients who are unable to consent. This provision reiterates the standard set by Law No. 20,584, which has very negative consequences for research and medical practice (Ruiz et al., 2021). The explicit prohibition on the use of Brain-Computer Interfaces (BCIs) on patients who cannot consent would certainly hinder the most significant clinical application of these methodologies: their use to facilitate communication in patients who are unable to interact with their external environment due to severe neurological conditions. A locked-in patient would not be able to give their consent. If this were to be approved, Chile would become the only country where such patients would not be able to access BCIs. This is something that clearly should not be imitated by the international community.

5 Conclusion

This chapter has critically examined the first attempt in the world to integrate neuroprotective safeguard in a legal framework. As seen, the discussion about how to include Neurorights is at the intersection of a number of philosophical and legal debates. If it is finally shown that current legal frameworks can deal with the type of threats posited by neurotechnological applications the whole Neurorights crusade would become meaningless. However, there seems to be good reason to believe that a number of issues regarding the relationship between neural data and the person cannot be properly covered by existing laws. In addition, explicit inclusion of neuro-protective laws could serve as a political act in favors of the protection of the human person in all her dimensions. However, as we can see in the analysis we have offered, a number of issues remain open. In this chapter we have described a specificized in detail some of the main elements of the process by which the Chilean constitution was reformed. We hope that the international community could draw some conceptual and practical lessons from this process, and, with this, inform future policy-making all around world considering the positive and negative aspects of the process we have witnessed.

References

A Adams S Albin K Amunts T Asakawa A Bernard JG Bjaalie K Chakli JO Deshler Y Koninck De CJ Ebell G Egan ME Hale M Häusser S-J Jeong J Illes L Lanyon P Li Y Li P Magistretti J Zheng 2020 International brain initiative: An innovative framework for coordinated global brain research efforts Neuron 105 2 212 216 https://doi.org/10.1016/j.neuron.2020.01.002

Agar, N. (2014). *Truly human enhancement: A philosophical defense of limits.* MIT Press.

AP Alivisatos M Chun GM Church K Deisseroth JP Donoghue RJ Greenspan PL McEuen ML Roukes TJ Sejnowski PS Weiss R Yuste 2013 The brain activity map Science 339 6125 1284 1285 https://doi.org/10.1126/science.1236939

D Borbón L Borbón 2021 A critical perspective on NeuroRights: Comments regarding ethics and law Frontiers in Human Neuroscience 15 703121https://doi.org/10.3389/fnhum.2021.703121

JC Bublitz 2022 Novel Neurorights: From nonsense to substance Neuroethics 15 1 7 https://doi.org/10.1007/s12152-022-09481-3

Bublitz, J.-C. (2013). My mind is mine!? Cognitive liberty as a legal concept. In E. Hildt & A. G. Franke (Eds.), *Cognitive enhancement* (Vol. 1, pp. 233–264). Springer Netherlands. https://doi.org/10.1007/978-94-007-6253-4_19

M Decker T Fleischer 2008 Contacting the brain—Aspects of a technology assessment of neural implants Biotechnology Journal 3 12 1502 1510 https://doi.org/10.1002/biot.200800225

F Gilbert I Russo 2024 Mind-reading in AI and neurotechnology: Evaluating claims, hype, and ethical implications for neurorights AI and Ethics 4 3 855 872 https://doi.org/10.1007/s43681-024-00514-6

A Gilead 2015 Can brain imaging breach our mental privacy? Review of Philosophy and Psychology 6 2 275 291 https://doi.org/10.1007/s13164-014-0226-z

M Ienca R Andorno 2017 Towards new human rights in the age of neuroscience and neurotechnology Life Sciences, Society and Policy 13 1 5 https://doi.org/10.1186/s40504-017-0050-1

Kloss, C. S. (2017). *El derecho administrativo y la protección de las personas.* Ediciones Universidad Católica

W Koroshetz J Gordon A Adams A Beckel-Mitchener J Churchill G Farber M Freund J Gnadt NS Hsu N Langhals S Lisanby G Liu GCY Peng K Ramos M Steinmetz E Talley S White 2018 The state of the NIH BRAIN initiative The Journal of Neuroscience 38 29 6427 6438 https://doi.org/10.1523/JNEUROSCI.3174-17.2018

López Silva, P., & Madrid, R. (2022). Acerca de la protección constitucional de los neuroderechos: La innovación chilena. *Prudentia Iuris, 94*, 39–68. https://doi.org/10.46553/prudentia.94.2022.pp.39-68

López-Silva, P. (2021). Mejoras cognitivas y el problema de la inequidad: Entre bioconservadurismo y bioliberalismo. *Persona y Derecho*, 183–206. https://doi.org/10.15581/011.84.009

López-Silva, P. (2022a). The concept of mind in the neuroprotection debate. In P. López-Silva & L. Valera (Eds.), *Protecting the mind* (Vol. 49, pp. 9–18). Springer International Publishing. https://doi.org/10.1007/978-3-030-94032-4_2

P López-Silva R Madrid 2021 Sobre la conveniencia de incluir los neuroderechos en la Constitución o en la ley Revista Chilena De Derecho y Tecnología 10 1 53 https://doi.org/10.5354/0719-2584.2021.56317

López-Silva, P., & Valera, L. (Eds.). (2022). *Protecting the mind: Challenges in law, neuroprotection, and neurorights* (Vol. 49). Springer International Publishing. https://doi.org/10.1007/978-3-030-94032-4

P López-Silva A Wajnerman-Paz F Molnar-Gabor 2024 Neurotechnological applications and the protection of mental privacy: An assessment of risks Neuroethics 17 2 31 https://doi.org/10.1007/s12152-024-09565-2

Madrid, R., Widow, J.-L., & Valera, L. (2021). Prólogo. ¿Humanidad sin fronteras? Transhumanismo, Posthumanismo y Derecho. *Persona y Derecho*, 7–16. https://doi.org/10.15581/011.84.001

J Ngai 2024 BRAIN @ 10: A decade of innovation Neuron 112 18 3003 3006 https://doi.org/10.1016/j.neuron.2024.09.007

Nogueira, H. (2005). Aspectos de una Teoría de los Derechos Fundamentales: La Delimitación, Regulación, Garantías y Limitaciones de los Derechos Fundamentales. *Ius et Praxis, 11*(2). https://doi.org/10.4067/S0718-00122005000200002

I Persson J Savulescu 2008 The perils of cognitive enhancement and the urgent imperative to enhance the moral character of humanity Journal of Applied Philosophy 25 3 162 177 https://doi.org/10.1111/j.1468-5930.2008.00410.x

Ratoff, W. (2024). The right to mental autonomy: Its nature and scope. *Journal of Ethics and Social Philosophy, 27*(2). https://doi.org/10.26556/jesp.v27i2.2907

Ruiz, S., Ramos Vergara, P., Concha, R., Altermatt, F., von Bernhardi, R., Cuello, M., Godoy, J., Valera, L., Araya, P., Conde, E., Toro, P., & Caneo, C. (2021). Efectos negativos en la investigación y el quehacer médico en Chile de la Ley 20.584 y la Ley de Neuroderechos en discusión: La urgente necesidad de aprender de nuestros errores. *Revista Médica de Chile, 149*(3). https://www.revistamedicadechile.cl/index.php/rmedica/article/view/9120

D Susser LY Cabrera 2024 Brain data in context: Are new rights the way to mental and brain privacy? AJOB Neuroscience 15 2 122 133 https://doi.org/10.1080/21507740.2023.2188275

United Nations. (2010). *International convention for the protection of all persons from enforced disappearance.* OHCHR. https://www.ohchr.org/en/instruments-mechanisms/instruments/international-convention-protection-all-persons-enforced

P Wolpe 2009 Is my mind mine? Neuroethics and brain imaging V Ravitsky A Fiester A Caplan Eds The Penn center guide to bioethics Springer Publishing Co 86 93

R Yuste S Goering BAY Arcas G Bi JM Carmena A Carter JJ Fins P Friesen J Gallant JE Huggins J Illes P Kellmeyer E Klein A Marblestone C Mitchell E Parens M Pham A Rubel N Sadato J Wolpaw 2017 Four ethical priorities for neurotechnologies and AI Nature 551 7679 159 163 https://doi.org/10.1038/551159a

Zagrebelsky, G. (2018). *El derecho dúctil: Ley, derechos, justicia* (1st edn). Trotta, Editorial S.A.

Zúñiga-Fajuri, A., Miranda, L. V., Miralles, D. Z., & Venegas, R. S. (2021). Neurorights in Chile: Between neuroscience and legal science. In *Developments in neuroethics and bioethics* (Vol. 4, pp. 165–179). Elsevier. https://doi.org/10.1016/bs.dnb.2021.06.001

A Situated Approach to Neurorights Legislation in Chile

Jorge Ignacio Fuentes, Paulina Ramos Vergara, Abel Wajnerman Paz, Pablo López-Silva, and Florencia Álamos

Abstract Some have advocated for creating new fundamental neurorigths with universal validity such that national legislations worldwide should adapt to enforce them. Others have brought into question such a movement warning against rights inflation. We argue that we should avoid introducing new fundamental rights in any legal framework whenever possible. We call this the situated approach since whether it is possible to protect the mind without introducing new rights is a question that should be asked in each legal framework separately. We show that in the case of Chile, that can be done, and we propose how. Specifically, we provide a critical assessment of local regulatory proposals and show that, as a result of the local legislative debate, they have evolved from a universalist stance towards a much more situated view. However, we suggest that in their current state, they still are insufficiently grounded in the Chilean legal system, which we regard as providing enough protection against neurotechnological harm in most relevant cases.

Keywords Neurorights · Neurotechnology regulation · Neuroethics · Mental integrity · Mental privacy

J. I. Fuentes (✉) · P. Ramos Vergara · F. Álamos
Centro de Bioética UC, Facultad de Medicina, Pontificia Universidad Católica de Chile, Santiago, Chile
e-mail: jfuenm@uc.cl

A. Wajnerman Paz
Instituto de Éticas Aplicadas (IEA), Pontificia Universidad Católica de Chile, Santiago, Chile

P. López-Silva
School of Psychology and Research Centre in Cognition and Language Development, Universidad de Valparaíso, Valparaíso, Chile

Millenium Institute for Research in Depression and Personality (MIDAP-Chile), Valparaíso Institute for Complex Systems, Valparaíso, Chile

Collaborating Centre for Values-Based Practice in Health and Social Care, University of Oxford, Oxford, UK

P. López-Silva (ed.), *Contextualizing Neuroprotection: Latin American Perspectives on the Impact of Neurotechnological Development in Life and Society*, Ethics of Science and Technology Assessment 54, https://doi.org/10.1007/978-3-031-96055-0_12

1 Introduction

A central debate concerning neurotechnology regulation is the very question of whether new legal instruments to protect people from potential harm are really needed (Ligthart et al., 2023b). Some suggest the creation of new human rights with universal validity, namely, neurorights. We shall call this position the 'universalist approach' (e.g. Göering et al. 2021; Ienca & Andorno, 2017; Ligthart et al., 2023a; Yuste et al., 2021). In contrast, others claim that new challenges and threats posed by neurotechnology use should be confronted with current local legal frameworks, at least to the degree that is possible. In other words, new fundamental rights must only be inserted into a legal framework if currently existing instruments seem entirely unfit to protect people from a particular threat. We shall call this view the situated approach (e.g. Bublitz, 2022).

This chapter argues for a situated approach to neurotechnologies' legal regulation by focusing on the particularities of the Chilean legal framework and the ongoing legislative debate regarding neurotechnologies regulation. Firstly, in Sect. 2, we clarify the main aspects of this legislative debate. Secondly, in Sect. 3, we aim to explain why protecting the human mind and its integrity is crucial and how neurotechnologies interact with the mental. We also explain why drugs are not neurotechnologies. Thirdly, in Sect. 4, we propose how current instruments in Chile might be used to protect people from neurotechnological harm without the need to introduce more fundamental rights than the one to psychological or mental integrity. Finally, we focus on two more specific debates surrounding the Chilean attempt at introducing neurorights to the Chilean legal framework: (i) the debate about mental privacy as a component of relational identity and autonomy (Wajnerman 2024b), and (ii) the debate concerning neural information and data and the processing of the former to infer the latter.

2 The Legislative Debate on (Neuro)rights in Chile

The Chilean debate has moved from a more universalist approach at the beginning of the process to a more situated approach in the current state of the debate. After much discussion, those involved started to weigh the consequences for medical research and treatment development of legally restricting innovation, especially for those with medical conditions that are more demanding of new medical therapies. In addition, there began to be a more nuanced assessment of specific risks that now appeared as more speculative than actual threats. The paradigmatic case was two articles initially intended to protect people from neurotechnological manipulation at the very moment of exercising their political right to vote. No currently known neurotechnology has such a power.

Regarding this last claim, it could be argued that the neurorights initiatives are inspired by previous "genie in the bottle" tech developments, according to which potentially harmful technologies should be regulated before they are created to prevent "the genie" from escaping.[1] Therefore, the current nonexistence of such technologies cannot work as an objection against universal neurorights proposals. However, preventing risks may involve substantial social costs, which is most often the case for technologies with medical applications (see Sect. 4). Limiting the development and application of neurotechnologies can potentially harm the interests of people whose quality of life could be improved through them (cf. Fins, 2022). Therefore, the proponents of the "genie in the bottle" approach have the burden of proof. They need to show (a) that the risks of these technologies are severe, (b) that existing legal systems cannot manage these risks, and (c) that they are probable enough to override the potential benefits of their development.

A caveat is necessary at this point. The situated approach we are to argue for here does not take that (neuro)rights are not needed nor essential in Chile but rather that they should not be understood as new rights in a literal sense. By contrast, we should legally understand them as adjustments to our current law instruments to protect people's mental integrity (as conceptualized in Wajnerman et al. 2024), and human agency and autonomy—personal identity concerns are to be subsumed under mental integrity ones, and mental privacy concerns under agency and autonomy ones.

The Chilean legislative debate started on October 7th, 2020, with the presentation of a constitutional bill to amend the first section of Article 19 (Boletín 13827-19) and a law bill (13828-19). We start by addressing the former.

The constitutional debate began with the presentation of a proposal to include the following subsection in section 1 of Article 19 of the fundamental law: "The physical and mental integrity allows people to fully enjoy their individual identity and freedom. No authority or individual can, by using any technological means, augment, diminish, or disturb individuals' integrity *without proper consent*. Only the common law is able to establish the requirements to limit this right and the requirements for such a consent." (Our translation and italics).

A crucial element of the constitutional debate was whether such a consent was in place. The reason to oppose it—defended in front of the Joint Commission of Deputies and Senators on September, 6th 2021, by professors Paulina Ramos Vergara and Carlos Amunategui—is that fundamental rights are inalienable and, thus, no individual could give up them (cf. Ruiz et al., 2024).

After this constitutional debate concluded, a different version of the subsection was approved and is now part of our fundamental chart. This new version recognizes the value of scientific inquiry—for the first time in Chile's constitutional history—and emphasizes that the ends of scientific inquiry are to be framed in terms of their service to people and removes the controversial claim regarding consent. Thus, it dictates:

Scientific and technological development will be at people's service and carried with respect for life and physical and psychological integrity. The common law will regulate the requirements, conditions, and restrictions for their use in persons, owing

[1] We thank an anonymous reviewer for this observation.

to specially safeguard the brain's activity as well as the information originating as its result.

In what follows until the end of the article, we move on to the common law bill on the Chilean neurorights debate. This bill is still under discussion and modification and, as such, has yet to be settled and approved. For this reason, our proposal has the additional goal of aspiring to contribute to the legislative debate to move forward.

This common law bill aimed to regulate (1) the use of neurotechnologies; (2) the specifics of their use in three different contexts, namely, therapy, scientific research, and commercial or consumer-oriented; (3) the accordance of neurotechnologies' deployment with fundamental rights established in our fundamental law; and finally, (4) a special regime of legal responsibility that includes criminal law sanctions (see Sect. 4).

As a condition, the law must define what we refer to when speaking of neurotechnologies to be non-ambiguous regarding its scope. A more than plausible requirement is that the new bill does not affect devices or items previously regulated with success in their applications. In this regard, pharmaceutical products or drugs should not be included in its range. However, as they also sometimes directly interact with the brain's functions, it takes some conceptual work to exclude them, and this task was essential in the Chilean debate (see Sect. 3.2).

3 Defining the Scope of the Debate

In this section, we would like to articulate what we regard as the critical dimensions of the debate on neurotechnology regulation. Firstly, we need to specify what we are protecting from neurotechnology misuse, namely, which aspects of the mind and the brain are ethically and legally relevant. Second, we need to characterize what we are protecting the mind from. This requires specifying the extension of the term 'neurotechnology' and distinguishing this kind of device from others that can also affect the mind and the brain. Finally, we need to explain how we are protecting the mind from these technologies. Specifically, a critical challenge is to find a balance between under and over-regulation.

3.1 What We Should Protect: Mental Integrity and Agency as Part of the Core of Human Dignity

Regarding the aspects of the mind that should be protected, Law Bill 13828-19 is very clear. Its title is "Bill on the protection of neurorights and mental integrity, and the development of research and neurotechnologies". Thus, it seems that mental integrity is the main focus of concern. Furthermore, this is emphasized in its first article:

> Article 1: The purpose of this law is to define the requirements, conditions, and restrictions for the use of neurotechnologies in clinical applications and scientific research, in order to protect the life and physical and *psychological integrity* of persons (our italics).

What is psychological or mental integrity? Articles 3 and 4 of The European Union Charter of Fundamental Rights are sometimes taken as articulating a positive and a negative dimension of the right to mental integrity. Its negative dimension consists of protection against torture and inhuman or degrading treatment or punishment, which includes purely psychological forms of torture and punishment. Most recent discussions of this right have been focused on this dimension. Specifically, neuroethicists have been trying to widen the scope of the interventions that could harm the mind and the brain, including, for instance, neurotechnological interventions that may not qualify as torture or punishment (e.g., Douglas & Forsberg, 2021; Ienca & Andorno, 2017; Lavazza, 2018). However, mental integrity is often also regarded as a positive right to mental health, a right that guarantees people with mental conditions access to mental health systems and psychiatric treatment. Recently, Wajnerman Paz et al. (2024) offered an approach in which mental health is a central feature that defines mental integrity.

Bill 13828-19 also seems to consider mental health a key concern. For instance, Article 7 states that all neurotechnologies should be registered by the Instituto de Salud Pública (ISP, *Institute of Public Health*)—our equivalent to the US's Food and Drug Administration (FDA) (see Sect. 4). This means that the effectiveness and safety of neurotechnologies should be focused on potential (positive and negative) effects on the health of users. Crucially, Article 8 states that The Institute of Public Health may assign a qualification to each registered neurotechnology "according to their impact on people's health."

Another aspect often associated with mental integrity in recent neurorights and neuroethics discussions is agency and autonomy. The mentioned reformulations of the right to mental integrity present it as a right against any non-consented intervention on a person's mental and brain states and processes. This means that a person should have control over third-party access and influence over her internal processes. This dimension of mental integrity can be regarded as captured, for instance, by Article 5, which specifies the necessary conditions for informed consent in a neurotechnology user. It demands that informed consent can only be obtained after explaining to a person the nature or features of the neurotechnology to be used, its expected benefits, known risks, the feasibility of reversing its effects, the life-cycle of the device, and its conditions for maintenance.

It should be noted that this negative right is different from the positive right to consent to any neurotechnological intervention, regardless of the effects it may have on the mind. This right, sometimes characterized as part of the right to cognitive liberty (Bublitz, 2013), has been discussed during the drafting of the bill. However, it was not incorporated into its final version because of the reason mentioned above, namely that fundamental rights are inalienable and therefore we cannot consent to interventions that may undermine them. For instance, if a neurotechnological device is capable of controlling our actions in a way that is inconsistent with agency, autonomy, and/or freedom of thought, its use should be prohibited regardless of consent status.

Based on these considerations, we conclude that key aspects of the mind to be protected, namely the health and agency dimensions of mental integrity, are well characterized by the bill.

3.2 *What We Should Protect Against: Problems with the Definition of Neurotechnology and Why It Should Exclude Pharmaceuticals*

Section two described that the common law bill aims to regulate the use of neurotechnology, making it imperative to establish a precise definition of this term. This, at first sight, seemingly straightforward task, however, proved to be intricate, requiring the alignment of criteria between the political sphere and the scientific community in Chile.

Does the word "neurotechnology" encompass tools that allow access to nervous system data, such as neuroimaging or electrophysiological tools like electroencephalography (EEG) or electrocorticography (ECoG)? Or does it extend to techniques that enable us to modulate the nervous system's activity, such as Deep Brain Stimulation (DBS) or Transcraneal Magnetic Stimulation (TMS)? Does this definition pertain only to the central nervous system, or does it also include the peripheral nervous system? Furthermore, does neuromodulation include drugs?

UNESCO defines neurotechnologies as devices and procedures used to access, monitor, investigate, assess, manipulate, and/or emulate the structure and function of the neural systems of animals or human beings (UNESCO, 2023). Under this definition, both reading and modulation methods fall under the category. However, it is crucial to understand the distinct nature of both, as different ethical concerns emerge from each one (Ligthart et al., 2023a, 2023b; Ratto Trabucco, 2023). While neuroimaging provides a 'snapshot' of neural data and plays a 'passive' role concerning the person involved,[2] neuromodulation actively intervenes in neural circuits, directly modifying the person's behavior and affecting agency (Ruiz et al., 2024).

Neurotechnologies can interact with the central nervous system (e.g., DBS or spinal cord stimulation) and the peripheral nervous system (e.g., vagus nerve stimulation). However, drugs should be excluded from the definition of neurotechnology for at least two reasons: the first one is ontological, and the other one is legal.

The ontological reason is that neurotechnologies belong to a different class of physical structure. Neurotechnologies are functional mechanisms (e.g. Fuentes,

---

[2] However, some neuroethicists have claimed (e.g. Gilbert et al., 2023; Heerminsk, 2015; Heinrichs, 2021; Fuentes and Wajnerman Paz forthcoming) that if actions are mediated by neurotechnology, as in brain-computer interface (BCI) mediated action, there is some sort of shared agency such that the device can, in some way, be integrated into the person given certain conditions.

2023; Garson, 2013; Maley & Piccinini, 2018), meaning that they can be decomposed into non-equivalent parts (Wimsatt, 1986) that perform different causal roles and that they can fail to pursue. If you split an X mg drug into two pieces, you will have two pieces of drug[3] of the same kind as the original of Y mg and Z mg such that $X = Y + Z$. If you split a neurotechnology, you will have different parts that are supposed to contribute to the overall function of the device in different ways.[4] For instance, cochlear implants (more on this example in section 4) have microphones, antennas, receptors, cables, a simple computational device to classify sound waves into components, and stimulation electrodes attached to a person's cochlear nerve. They all do different things. Why is this ontological reason normatively relevant? While by the use of drugs, we can intervene in neurocognitive mechanisms along a single dimension (i.e., the dose supplied), functional mechanisms offer the possibility of a multiple-dimensional manipulation (Craver, 2007; Piccinini, 2020; Woodward, 2004). For instance, by maliciously manipulating a cochlear device, we could make its user hear the right things but with less or more intensity than is supposed. Still, we could also manipulate the pitch by affecting the frequencies algorithm and maybe even, with a more complex intervention, the semantic content of a heard message.

Furthermore, in medical contexts, neurotechnologies provide more fine-grained and effective interventions. For instance, DBS trials have proved that this kind of device is highly precise regarding the targeted structure and more efficient than drugs for some conditions, such as treatment-resistant depression (Wajnerman 2024a).

The legal reason why drugs should be excluded is that, at least in our country (Chile), there is already a well-functioning regulation and a clear definition for them. This definition states:

> A pharmaceutical product or medicine shall be understood as any natural, biological, synthetic substance or mixtures thereof, originated through synthesis or chemical, biological or biotechnological processes, which is intended for people for prevention, diagnosis, attenuation, treatment or cure of diseases or their symptoms or regulation of their particular physiological systems or states, including in this concept the elements that accompany its presentation and that are intended for its administration.

This definition includes both prescription and non-prescription drugs. Drugs already must be authorized by the ISP before being marketed in Chile, guaranteeing that they meet the necessary quality standards, safety, and efficacy. Therefore, to avoid confusion, the definition of neurotechnology in the bill should be adjusted to state the exclusion of drugs explicitly.

---

[3] Of course, you can split a drug's active ingredient molecule by breaking any of its chemical bonds. However, the relevant level of mechanistic analysis is neurocognitive mechanisms. We are interested in how drugs affect the dynamics that constitute psychological capacities or brain development rather than on the molecular level of organization of matter. We thank an anonymous reviewer for this observation.

[4] A further ontological difference concerns the kind of interaction. Drugs perform their function in such a way as to be assimilated by the human body. In other words, their (intended) effects progressively disappear from the body as they are metabolized.

Also, being transparent about the scope and limitations of the concept of neurotechnology helps us clarify and strengthen normative regulation. By understanding what neurotechnology is, we can ensure that the norms are grounded in actual aspects of such devices and not solely based on speculative assumptions. Simply put, this conceptual knowledge will ultimately contribute to developing more effective and informed regulatory frameworks.

4 Prevention of and Responsibility for Harm

Guaranteeing a right to protection always results in restricting freedom for the protected person and/or others. We must carefully assess this tradeoff between freedom and protection on a case-by-case basis. In its current state, the Chilean bill creates a specially harsh regime of responsibility, which we are to argue against in the current section. This regime has three elements. (I) One is the characterization of civil liability—i.e., legal responsibility—for damages in using neurotechnologies as solidary and objective—we further explain below what this means in our system. (II) The second is the establishment of criminal responsibility in some cases. (III) The third is that neurotechnologies must be registered at the ISP regardless of whether their intended use is therapy, scientific research, or consumer-oriented.

Some points are worth noting here. First, every other medical device in Chile must comply with (III) but is nevertheless subject to a subjective liability regime. Second, no harm produced by using a medical item is currently subject to criminal law in our system, and (II) intends to change this. As a consequence of these first two points, an unprecedentedly robust regime that goes, as the former points show, far beyond other medical items is intended to be established. Third, anyone wanting to register a medical item at ISP must explicitly state its intended uses and their forbidden uses and provide evidence of it being sufficiently tested for its potential adverse effects in the form of medical testing studies before this registration. Moreover, no other consumer-oriented kind of device currently requires such a standard. Consequently, an unprecedentedly robust regime for a consumer-oriented device is also intended to be established. While this second consequence is highly desirable, given the importance of the human brain's activity, the first one is problematic.

The problem with such high standards is that when conjoined with the already severe ones for medical devices in our legislation, they create a strong barrier that precludes scientists and entrepreneurs from pursuing research and innovation. This, in turn, affects the availability of new medical treatments for those who need them the most.

Let us address civil responsibility. The proposed legal regime would establish solidary and objective responsibility. Article 9 states: "The producer, the provider, and anyone who supplies neurotechnologies to a consumer, will be responsible, solidary and objectively for the material and moral damage that they could cause." Let us analyze each of these qualificative roles—at least in the Chilean legal framework and similar enough ones. Solidary responsibility means they—producer, provider, and anyone who supplies—share equal liability in case of damage so that the affected user can legally sue any of them depending on their geographical proximity and economic capacity. Its opposite is individual responsibility that can be attached to a specific actor.

Objective responsibility is the opposite of subjective responsibility. The latter is attributed when a judge considers that the accused person acted on intentional wrongdoing or without taking all reasonable precautions—on Spanish *dolo* from the Latin term *dolus* (deceit, trickery, or bait). Thus, objective responsibility is attributed regardless of blameworthiness but solely based on actual damage. This objective responsibility regime is exceptional in our legal framework (Díaz Tolosa, 2007). Furthermore, this regime is not intended to cover medical items—be they drugs or devices—as they have their own regime. Article 10 further establishes prison sentences in some cases, moving from civil to criminal law.

While we agree that the human mind and its neural realizers are worthy of the highest protection, we need to balance the consequences of conjoining the high standards already in place for medical items with this solidary and objective liability, plus the criminal sanctions proposed. We need to think of these consequences, especially in connection with the development of medical treatments for those most deserving.

In what follows, we articulate reasons to think (a) that such consequences are not desirable and that (b) past-generation neurotechnologies were implemented in our country long ago without legal complications, so requirement (III) plus subjective responsibility has been shown to be sufficient in practice. Based on the latter, we ground our belief that much of the legal anxiety regarding neurotechnologies lacks an evidence-based foundation.

Let us start by describing potential negative consequences. A noticeable legal precedent in Chile is Law 20.584, promulgated in 2012. As qualitatively shown by Ruiz et al. (2021), this law had a blowing impact on medical research in Chile for the development of treatments for conditions such as depression, bipolar disorder, schizophrenia, Parkinson's disease, Alzheimer's disease, and multiple sclerosis. The reason is that the mentioned dictated that "No person with a psychic or intellectual disability who is unable to state her free will can participate in scientific research." It is very likely that conditions (I), (II), and (III), by being more robust than this historical example, conjointly cause an equal or higher impact on research and innovation. Furthermore, by imposing such a rigid civil liability, suppliers are being pushed to pay related insurance to cover the associated risks that could result in higher prices and lower accessibility.

We now move on to advance the idea that stimulation neurotechnologies are only a partial novelty in our country. Moreover, a specific stimulation neurotechnology, the cochlear implant, has been applied in Chile since 1983,[5] and our legal system has proved sufficient in dealing with its implementation. We expect this exemplar case to reduce legal disquietude regarding other stimulation technologies, at least

[5] The first cochlear implant was implemented in Chile in 1983 in the Hospital Militar by MDs Marco Goycoolea Vial and Gumaro Martínez Concha and the anesthesiologist Yolanda Gho Barba. However, these implants became a more common medical treatment during the 90s.

those neurotechnologies similar in nature, i.e., those aimed at restoring capacities in persons with perceptual disabilities—always provided patients involved consent to such treatments.

Cochlear implants aim to restore hearing perceptual capacity. They translate sound waves into a code amenable to the cochlear nerve via a relatively simple algorithm that classifies and separates these waves into their components and stimulates the relevant parts of the nerve with electrodes that carry the analog representations (Maley, 2023) of each of those components (e.g. Dorman & Wilson, 2004). Although neurotechnologies under development are progressively more complicated, there is no principled reason to believe that other perception-restoring neurotechnologies are different in kind. As legislation has worked well in these devices, we must follow folk wisdom in the prescription: *If it ain't broke, don't fix it.*

In summary, the combination of (III) and the current regime of subjective responsibility[6] for medical items seems enough to protect people from neurotechnological harm. Any more demanding protection could be detrimental to those who need medical innovation. However, this combination puts an unparalleled high standard for a consumer-oriented device, which we consider positive given the importance of the human mind.[7]

5 Neural Data and Mental Privacy: Critical Comments and Open Challenges

Article 3-a of the bill defines neural data as 'any information derived from the activity of the cells that compose the central or peripheral nervous system of a person.' The bill also defines biological data as "any information relating to the biological or genetic characteristics, properties, compositions or biochemical or physiological activity attributable to a person". These definitions seem to capture key aspects of the connection between neural and biological data. However, as such, these definitions seem to leave out important ways in which neural data has also been defined.

Firstly, neural data includes more than information about neural activity. There are at least two additional kinds of data that should have the same label. Information about the structural properties of the brain is also a basic neural data category. For instance, cytoarchitecture allows anatomical differentiation of areas in the cerebral cortex based on changes in different parameters of neuronal organization, such as density (i.e., the number of cells per unit volume), size or type of cells, the relative thickness of the layers of the cortex (one layer may thicken at the expense of others

[6] Solidary responsibility means that any of the mentioned bears equal responsibility. The current regime, by contrast, allows the judges to decide on a case-by-case basis who among them is responsible and to which degree.

[7] Maslen et al. (2014) propose a similar model for enhancement devices. However, we do not distinguish here between consumer-oriented devices for enhancement and those for entertainment or any other purposes.

or vice versa) and increase or decrease in the thickness of the cortex as a whole. Here, it is important to note that information about structural damage to the brain is often used to understand psychopathology and ground judgments of legal capacity.[8] In the same way, misuse of this type of information (that is also neural data) could damage people, in the same way in which neural data—as defined in the bill—could be misused. Problematically, it is not entirely clear how the current version of the bill explicitly protects this type of neural data.

Another kind of information that is also considered neural data relates to the functional properties of brain structures. Functional and physiological properties are often confused in the literature. For instance, in network neuroscience, 'functional connectivity' is often used to refer to the correlation between the physiological activity of different brain structures (as opposed to their 'structural connectivity'). However, functional properties are often different from physiology in that they refer to the 'goal' according to which structural and physiological properties are organized. For instance, in design explanations of neural mechanisms, these refer to the design variables the mechanism needs to optimize (e.g., metabolic cost, speed, and accuracy of signal transmission). As much as structural properties, functional properties are critical for identifying pathological conditions if we understand disease as a deviation from normal function (Boorse, 2014). These deviations may involve both structural and/or physiological properties. Knowledge of this information could be used or misused in different situations, and it is not clear how the current bill on neurorights covers it.

Thirdly, neural data is often considered to include any proxy of the nervous system's activity. For example, functional magnetic resonance techniques that measure hemodynamic and metabolic changes in the brain that accompany neuronal activation (David et al., 1994) are also considered relevant.

Fourthly, even if we endorse the definition of biological data offered in the bill, there is a key feature that makes neural data fundamentally different from it. Many (or most) neural mechanisms are often characterized as 'cognitive' (or 'neurocognitive') because they are dedicated to decoding, processing, storing, and/or transmitting information. As we mentioned before, the structural and physiological organization of neural mechanisms can often be explained as an optimization strategy for efficiently performing different informational tasks (Chirimuuta, 2014; Fuentes, 2023; Sterling & Laughlin, 2015; Wajnerman Paz, 2017). This means that the brain also produces, manipulates, and transmits data about external and internal variables (including the brain itself). Thus, neural data should include not only information about the brain that is gathered through 'reading' neurotechnologies, such as EEG and MRI, but also information *in* the brain, *data embedded in neural code.* This means that there are some forms of neural data that cannot be distinguished from the

[8] In a famous case, Aaron Hernandez was sentenced in 2015 to life in prison for the murder of Odin Lloyd during 2013. After commiting suicide in 2017 an analysis of structural damage of his brain showed that he suffered from chronic traumatic encephalopathy, which could have explained his violent behavior.

brain itself, an issue that Ienca and Andorno (2017) characterized as "the inception problem".

A fifth issue that we would like to mention is related to the distinction between neural and mental data. It is clear that there are many kinds of information about the mind that cannot be characterized as neural data because they are derived from the analysis of information about behavior rather than about neural processes. In other cases, information that is considered 'mental' is considered to be irreducible to its physical components (López-Silva 2022a, 2022b). However, in the case where this information is inferred from neural data, even if its content is entirely about purely psychological properties, it should be regarded as neural data and protected as such. Some of the aspects that make neural data particularly sensitive are not related to information about neural properties but rather to its potential contribution to making reliable inferences about the subject's mental states.

Relatedly, it is worth noticing that these data allow us to infer mental contents but not their associated phenomenal qualities. The fact that they cannot nor will ever be able to do this has been called the hard problem of consciousness (Chalmers, 2010). Understanding this limitation is critical since reading-neurotechnologies do not give us access to the subjective aspects of mental representations. However, this difference does not preclude neurotechnologies from, despite this limitation, being able to modulate both.

Finally, we would like to comment on the practical implications of the target bill regarding the protection of neural data and mental privacy. Article 6 of the bill declares neural data to be treated as sensitive data under the terms of Law No. 19,628 on the Protection of Privacy. Neural data, then, may be considered biometric data under the terms of the just mentioned law if it allows the identification of a specific person. According to this law, specific inferences from biometric data such as genomic and neuronal data—are considered of a sensitive type.

The Chilean Congress just modified this law in August 2024, and the new version is close to being promulgated.[9] In Article 2, this update defines

> Sensitive personal data: we shall consider under such condition any personal data that refers to the physical or moral traits of persons or to facts or circumstances of their lives or their intimacy or that could reveal their ethnical belonging, their race, their political filiation, or filiation to a union, their socioeconomic situation, their ideological or philosophical beliefs, their religious beliefs, their health status, their biological human profile, their biometric data, and the information regarding their sexual activity, their sexual orientation, and their gender identity.

This law creates powerful economic sanctions for any company that does not guarantee the protection of their stored data. Furthermore, it mandates the creation of a new National Agency for Personal Data Protection whose sole purpose is to ensure that companies behave according to this law. Note that the concerns usually employed to sustain new neurorights' necessity are included in the Article cited, not just for neuronal data but for any kind of data.

---

[9] See https://www.gob.cl/noticias/ley-proteccion-datos-personales-aprobacion-eleva-estandar-derechos/.

Thus, through its connection to (the new version of) law 19,628, Law Bill 13828-19 determines that neural data may only be processed when authorized by the former law, in the event that it is necessary for the protection of a vital interest of the person or when the person gives express, free, and informed consent to it, and such consent must be recorded in writing. Law bill 13828-19 also demands that the person responsible for processing this type of data must have a system of risk management and evaluation of the impact of the processing on the rights of the data subject.

It is important to see how the current version of the bill discussed has moved away from the organic approach that characterized initial proposals. This approach proposed that:

> The collection, storage, treatment, and dissemination of neuronal data and the neuronal activity of individuals will comply with the provisions contained in Law No. 19.451 regarding transplantation and organ donation, as applicable, and the provisions of the respective health code.

We believe that this is a clear step forward toward the creation of efficient ways of protecting neural data, as the organic approach has been largely criticized over the last years (Bulitz, 2022; López-Silva et al., 2024). For space reasons, we cannot examine all these criticisms; however, one of the main practical consequences of the changes proposed in the law is that no new right to mental privacy is proposed or secured. In fact, no new special regulation is created as neural data is simply covered by already existing laws on sensitive data.

The modification in the wording of the law does not involve any conceptual modification either, as all that it does is make explicit that neural data fall into the sensitive data category. It could be argued that this leaves an important issue open. Recently, López-Silva et al. (2024) have argued that neural data would be different from other kinds of sensitive data. Firstly, from a biological point of view, neural data are causally closer to behavior than other kinds of sensitive data, such as genetic data. Secondly, neural data exhibits a higher temporal resolution than other types of sensitive data and, therefore, it provides a much powerful tool for manipulating behavior. Neural data can provide access not only to relatively stable psychological traits and dispositions but also to individual, dynamic, and temporally indexed mental states, events, and processes. Thirdly, neural data may be seen as providing more "direct" access to our mental processes and states than behavioral data, to which many psychological analytic tools are applied. When we obtain mental information from neural data, it seems that we are gathering this information "directly from its source" so to speak. Finally, as has already been mentioned, some kinds of neural data can be regarded not merely as information *about* the brain but rather as information *in* the brain.

Although the current version of the law bill does not seem to reflect the unique nature of neural data, this uniqueness does not necessarily entail that we need new regulations specifically tailored to them. It is possible that the kind of robust protection provided by the new national law on data protection is enough. Despite the special harms that could result from the misuse of neural data, it is not obvious that they could not be prevented by current (or upcoming) laws. However, we suggest a cautious case-by-case analysis of risks associated with neural data to confirm that more robust protection would not be needed.

6 Conclusion

In this chapter, we have addressed the central question in the neurorights debate: whether we need new legal instruments to protect people from potential neurotechnological harm from a situated perspective. Specifically, we exemplified a situated approach by assessing whether the Chilean legal system already has the tools to protect people from such harm.

Of course, one could mean different things by ‘new legal instruments.’ For instance, one could mean introducing new fundamental rights or simply new ordinary laws. Moreover, this last case can have two interpretations. A new law could be one that introduces restrictions or permissions that do not follow from or are not implied by any current regulation. However, a neurorights law could simply specify which and how already existing laws will apply to neurotechnologies. As we can see, Law Bill 13828-19 is a combination of the last two elements. It does not introduce new fundamental rights but rather specifies how current law will apply to neurotechnology (e.g., neural data) and proposes some new regulations (prevention from and responsibility for harm).

We have argued that the kind of integration of neural data to current privacy regulation (which is about to be promulgated) may be enough, although a more thorough risk assessment may be necessary. Regarding legal innovation, we suggested that the responsibility regime would be highly problematic given recent Chilean cases of over-regulation and the risk of undermining medical research. However, we agreed with the idea of applying to commercial neurotechnology the medical technology safety and control standards, which could be a way of legally acknowledging the special risks associated with these devices.

We regard our criticisms of the current version of bill 13828-19 simply as a continuation of the kind of development that the bill has had since its first version, that this, moving away from a generic or universalist view on neurorights to a situated stance in which concerns about neurotechnologies are progressively grounded on our local legal system, a process which may still leave room for some reasonable innovation.

References

Boorse, C. (2014). A second rebuttal on health. *Journal of Medicine and Philosophy, 39*(6), 683–724. https://doi.org/10.1093/jmp/jhu035

Bublitz, J. C. (2013). My mind is mine!? Cognitive liberty as a legal concept. In E. Hildt, & A. G. Franke (Eds.), *Cognitive enhancement* (Vol. 1, pp. 233–264). Springer Netherlands. https://doi.org/10.1007/978-94-007-6253-4_19

Bublitz, J. C. (2022). Novel neurorights: From nonsense to substance. *Neuroethics, 15*(1), 7. https://doi.org/10.1007/s12152-022-09481-3

Chalmers, D. J. (2010). *The character of consciousness.* Oxford University Press. https://doi.org/10.1093/acprof:oso/9780195311105.001.0001

Chirimuuta, M. (2014). Minimal models and canonical neural computations: The distinctness of computational explanation in neuroscience. *Synthese, 191*(2), 127–153. https://doi.org/10.1007/s11229-013-0369-y

Craver, C. F. (2007). *Explaining the brain.* Oxford University Press. https://doi.org/10.1093/acprof:oso/9780199299317.001.0001

David, A., Blamire, A., & Breiter, H. (1994). Functional magnetic resonance imaging: A new technique with implications for psychology and psychiatry. *British Journal of Psychiatry, 164*(1), 2–7. https://doi.org/10.1192/bjp.164.1.2

Díaz Tolosa, R. I. (2007). Responsabilidad objetiva en el ordenamiento jurídico chileno. *Revista de Derecho (Coquimbo), 14*(1), 79–112. https://doi.org/10.22199/S07189753.2007.0001.00004

Dorman, M. F., & Wilson, B. S. (2004). The design and function of cochlear implants. *American Scientist, 92*, 436–445. https://www.americanscientist.org/sites/americanscientist.org/files/2005414113245_306.pdf

Douglas, T., & Forsberg, L. (2021). Three rationales for a legal right to mental integrity. In S. Ligthart, D. Van Toor, T. Kooijmans, T. Douglas, & G. Meynen (Eds.), *Neurolaw* (pp. 179–201). Springer International Publishing. https://doi.org/10.1007/978-3-030-69277-3_8

Fins, J. J. (2022). The unintended consequences of Chile's neurorights constitutional reform: Moving beyond negative rights to capabilities. *Neuroethics, 15*(3), 26. https://doi.org/10.1007/s12152-022-09504-z

Fuentes, J. I. (2023). Efficient mechanisms. *Philosophical Psychology, 0*, 1–24. https://doi.org/10.1080/09515089.2023.2193216

Fuentes, J. I., & Wajnerman Paz, A. (forthcoming). El rol de la agencia en la identidad híbrida de los pacientes neurotecnológicos. *Revista de Filosofía.*

Garson, J. (2013). The functional sense of mechanism. *Philosophy of Science, 80*, 317–333. https://doi.org/10.1086/671173

Gilbert, F., Ienca, M., & Cook, M. (2023). How i became myself after merging with a computer: Does human-machine symbiosis raise human rights issues? *Brain Stimulation, 16*(3), 783–789. https://doi.org/10.1016/j.brs.2023.04.016

Goering, S., Klein, E., Specker Sullivan, L., Wexler, A., Agüera Y Arcas, B., Bi, G., Carmena, J. M., Fins, J. J., Friesen, P., Gallant, J., Huggins, J. E., Kellmeyer, P., Marblestone, A., Mitchell, C., Parens, E., Pham, M., Rubel, A., Sadato, N., Teicher, M., Yuste, R. (2021). Recommendations for responsible development and application of neurotechnologies. *Neuroethics, 14*(3), 365–386. https://doi.org/10.1007/s12152-021-09468-6

Heersmink, R. (2015). Dimensions of integration in embedded and extended cognitive systems. *Phenomenology and the Cognitive Sciences, 14*(3), 577–598. https://doi.org/10.1007/s11097-014-9355-1

Heinrichs, J. H. (2021). Neuroethics, cognitive technologies and the extended mind perspective. *Neuroethics, 14*(1), 59–72. https://doi.org/10.1007/s12152-018-9365-8

Ienca, M., & Andorno, R. (2017). Towards new human rights in the age of neuroscience and neurotechnology. *Life Sciences, Society and Policy, 13*(1), 5. https://doi.org/10.1186/s40504-017-0050-1

Lavazza, A. (2018). Freedom of thought and mental integrity: The moral requirements for any neural prosthesis. *Frontiers in Neuroscience, 12*, 82. https://doi.org/10.3389/fnins.2018.00082

Ligthart, S., Ienca, M., Meynen, G., Molnar-Gabor, F., Andorno, R., Bublitz, C., Catley, P., Claydon, L., Douglas, T., Farahany, N., Fins, J. J., Goering, S., Haselager, P., Jotterand, F., Lavazza, A., McCay, A., Wajnerman Paz, A., Rainey, S., Ryberg, J., & Kellmeyer, P. (2023a). Minding rights:

Mapping ethical and legal foundations of 'Neurorights.' *Cambridge Quarterly of Healthcare Ethics, 32*(4), 461–481. https://doi.org/10.1017/S0963180123000245
Ligthart, S., Bublitz, C., & Alegre, S. (2023b). Neurotechnology: We need new laws, not new rights. *Nature, 620*(7976), 950–950. https://doi.org/10.1038/d41586-023-02698-z
López-Silva, P. (2022a). The concept of mind in the neuroprotection debate. In P. López-Silva, & L. Valera (Eds.), *Protecting the mind* (Vol. 49, pp. 9–18). Springer International Publishing. https://doi.org/10.1007/978-3-030-94032-4_2
Lopez-Silva, P., & Madrid, R. (2022b). Protegiendo la mente: Un análisis al concepto de lo mental en la ley de neuroderechos. *Revista de Humanidades de Valparaíso, 20*, 101–117. https://doi.org/10.22370/rhv2022iss20pp101-117
López-Silva, P., Wajnerman-Paz, A., & Molnar-Gabor, F. (2024). Neurotechnological applications and the protection of mental privacy: An assessment of risks. *Neuroethics, 17*(2), 31. https://doi.org/10.1007/s12152-024-09565-2
Maley, C. J. (2023). Analogue computation and representation. *The British Journal for the Philosophy of Science, 74*(3), 739–769. https://doi.org/10.1086/715031
Maley, C., & Piccinini, G. (2018). A unified mechanistic account of teleological functions for psychology and neuroscience. In D. Kaplan (Ed.), *Explanation and integration in mind and brain science* (1st edn.). Oxford University Press. https://doi.org/10.1093/oso/9780199685509.003.0011
Maslen, H., Douglas, T., Cohen Kadosh, R., Levy, N., & Savulescu, J. (2014). The regulation of cognitive enhancement devices: Extending the medical model. *Journal of Law and the Biosciences, 1*(1), 68–93. https://doi.org/10.1093/jlb/lst003
Piccinini, G. (2020). *Neurocognitive mechanisms: Explaining biological cognition* (1st edn.). Oxford University Press. https://doi.org/10.1093/oso/9780198866282.001.0001
Ratto Trabucco, F. (2023). Neurorights between ethical and legal implications. *Cuadernos De Derecho Transnacional, 15*(1), 750–757. https://doi.org/10.20318/cdt.2023.7561
Ruiz, S., Ramos-Vergara, P., Concha, R., Altermatt, F., Von-Bernhardi, R., Cuello, M., Godoy, J., Valera, L., Araya, P., Conde, E., Toro, P., & Caneo, C. (2021). Efectos negativos en la investigación y el quehacer médico en Chile de la Ley 20.584 y la Ley de Neuroderechos en discusión: La urgente necesidad de aprender de nuestros errores. *Revista Médica de Chile, 149*(3), 439–446. https://doi.org/10.4067/s0034-98872021000300439
Ruiz, S., Valera, L., Ramos, P., & Sitaram, R. (2024). Neurorights in the constitution: From neurotechnology to ethics and politics. *Philosophical Transactions of the Royal Society B: Biological Sciences, 379*(1915), 20230098. https://doi.org/10.1098/rstb.2023.0098
Senate of Chile. (2020). Boletín 13827-19. *Constitutional reform.* See https://www.senado.cl/appsenado/templates/tramitacion/index.php?boletin_ini=13827-19
Senate of Chile. (2020). Boletín 13828-19. *Proyectos de Ley.* See https://www.camara.cl/legislacion/ProyectosDeLey/tramitacion.aspx?prmID=14385&prmBOLETIN=3828-19
Sterling, P., & Laughlin, S. (2015). *Principles of neural design.* The MIT Press. https://doi.org/10.7551/mitpress/9780262028707.001.0001
UNESCO. (2023). *Ethical issues of neurotechnology. IBC report series.* UNESCO. https://doi.org/10.54678/QNKB6229
Wajnerman Paz, A. (2017). Pluralistic mechanism. *Theoria, 32*(2), 161–175. https://doi.org/10.1387/theoria.16877
Wajnerman-Paz, A. (2024a). Identifying relational applications of deep brain stimulation for treatment resistant depression. *Review of Philosophy and Psychology, 15*(2), 499–521. https://doi.org/10.1007/s13164-023-00681-1
Wajnerman-Paz, A. (2024b). La privacidad mental como fundamento de la identidad y la autonomía relacional. *Revista de Humanidades de Valparaíso, 26*, 205–221. https://doi.org/10.22370/rhv2024iss26pp205-221
Wajnerman-Paz, A., Aboitiz, F., Álamos, F., & Ramos Vergara, P. (2024). A healthcare approach to mental integrity. *Journal of Medical Ethics, 50*(10), 664–669. https://doi.org/10.1136/jme-2023-109682

Wimsatt, W. C. (1986). Forms of aggregativity. In A. Donagan, A. N. Perovich, & M. V. Wedin (Eds.), *Human nature and natural knowledge* (Vol. 89, pp. 259–291). Springer Netherlands. https://doi.org/10.1007/978-94-009-5349-9_14

Woodward, J. (2004). *Making things happen.* Oxford University Press. https://doi.org/10.1093/0195155270.001.0001

Yuste, R., Genser, J., & Herrmann, S. (2021). It's time for neuro-rights. *Horiz. J. Int. Relat. Sust. Dev., 18*, 154–165.

Understanding Consent: Challenges in the Age of Neurotechnologies

Danielle Zaror-Miralles

Abstract Chilean regulation on neuro rights has served as proof for those who promote it that it is possible to regulate emerging technologies. However, this effort suffers from a series of weaknesses, where perhaps the lack of evidence is the most important. In this work I argue that in addition to being a redundant regulation because the risks it is supposed to protect are covered by other fundamental rights, it has been designed with a defective regulatory technique that builds all its strength based on consent. This enabling circumstance can only be delivered by healthy people and people with some level of deterioration in mental or brain activity, who benefit most from neuro technologies, are in practice not only left out of the progress of science but will also experience lower levels of access to the benefits guaranteed by virtue of the right of access to health.

Keywords Regulation · Consent · Legal capacity · Neurotechnologies

1 Preliminary Issues

Since the promoters of the regulation of neurotechnology's achieved their first victory with the reform of the Chilean Constitution in 2021:

The Constitution assures all people: scientific and technological development will be at the service of people and will be carried out with respect for life and physical and mental integrity. The law will regulate the requirements, conditions and restrictions for its use in people, especially safeguarding brain activity, as well as the information coming from it (art. 19 No. 1.).

The international debate on this matter has increased, has sharpened its positions, has demonstrated the problems of regulation and, above all, exposes that this is a discussion that is far from over.

D. Zaror-Miralles (✉)
Santiago, Chile
e-mail: dzaror@derecho.uchile.cl

P. López-Silva (ed.), *Contextualizing Neuroprotection: Latin American Perspectives on the Impact of Neurotechnological Development in Life and Society*, Ethics of Science and Technology Assessment 54, https://doi.org/10.1007/978-3-031-96055-0_13

As a result of this regulatory milestone, some questions have emerged whose answer seems to be much more pressing, among them, the way in which the consent of those who face the use of technologies that will be used at the brain level is expressed. To explain the above, it is necessary to recapitulate some contextual and conceptual issues that should be clear before going deeper into this matter.

Firstly, since Chile incorporated into its Constitution the first rule on special protection of brain activity and processed a bill that regulated the in the way neurotechnology's should be used in the country (a project that is still in process but has not moved since 2022, Boletín 13.828-29), there has been a real lobbying machine around the world in favor of its regulation, presenting the Chilean case in Seminars, Summits, International Organizations, and in general to anyone who has wanted to hear about the matter.

Some countries such as Mexico have included in their national agenda on artificial intelligence, among several laws, the enactment of one on neuro rights; Spain, for its part, included in 2021 in its charter of digital rights the conditions for the use of neurotechnology's, among which it states:

The conditions, limits and guarantees of implementation and use in people of neurotechnology's may be regulated by law in order to: (a) Guarantee the control of each person over their own identity. (b) Guarantee individual self-determination, sovereignty and freedom in decision-making. (c) Ensure the confidentiality and security of the data obtained or related to their brain processes and the full control and disposition of the same. (d) Regulate the use of human-machine interfaces that may affect physical or psychological integrity. (e) Ensure that decisions and processes based on neurotechnology's are not conditioned by the supply of incomplete, unwanted, unknown or biased data, programs or information (Digital Bill of Rights Secretaría de Estado de Digitalización e Inteligencia Artificial (2020)).

The majority of those who lend their ears to this kind of experiences are parliamentarians, who, as the theory of legislation has warned for many years, are part of a slow and inexperienced body in the matters they regulate (Blackstone bk I, sect. 1) and it is precisely for these reasons that an attempt has been made to orient this work towards an operationally rational model (Amado 1997, 309; Atienza 1989, 24). This lack of technical knowledge about the complexities of neurotechnology's and, in general, the sciences that it seeks to regulate, is the starting point of many of the problems that this legislative innovation is beginning to reveal.

Secondly, when we talk about neurotechnology's understood as a field of science and engineering in which the nervous system is interconnected with technical devices (Stieglitz, 2021, 4) or as the set of methods and instruments that allow a direct connection of technical components with the nervous system (Müller & Rotter 2017, 1), much of the discussion of its regulation is strongly promoted and influenced by certain fears that have been amplified by academics and businessmen and by fantasies with more roots in cinematography than in reality. There are several works that wander on the subject by presenting hypothetical scenarios instead of scientific evidence (CIOMS, 2024).

Without going any further, the same academic article that opens the discussion on neuro rights *Four ethical priorities for neurotechnologies and AI* in November 2017, states in several of its paragraph's hypothetical premises such as when it points out "Let us consider the following scenario. A paralyzed man participates in a clinical trial of a brain-computer interface (BCI). (…) This scenario is hypothetical, but it illustrates some of the challenges that society could face" or when it expresses:

It could take years or even decades until the BCI and other neurotechnologies become part of our daily lives, but technological advances indicate that we are on the way to a world in which it will be possible to decipher people's mental processes and directly manipulate the brain mechanisms that underlie their intentions, emotions and decisions (Yuste et al., 2017, 159–160).

In fact, if we review the statement of reasons for legislating that was presented in the Chilean Parliament for the two reforms, the constitutional and the legal, it is possible to notice that within the cases presented as reasons for legislating (Bulletin No. 13, 828) hypothetical cases are also presented, such as the Neuralink a business initiative that promised in 2020 that in one year, that is, 2021, it would be possible to "download thoughts". To date, the promise has not yet been fulfilled.

Another reason to justify the regulation was the mind mapping project of the University of California, Berkeley professor, Jack Gallant, who points out that the brains that are analyzed are similar, but not the same; one third of the patterns match, but two thirds present differences between individuals and in that sense the mapping, with computing techniques, can only yield an "average" result. Both legal initiatives are based on this scientific project to demonstrate that brain mapping is a fact that makes regulation urgent. However, it seems that being able to read and transcribe complex neuronal activity, as promised by the promoters of the idea, does not seem plausible either in the short or medium term, without prejudice to its value as predictors of general brain processes. Chilean scientists have pointed out that "it is necessary that there are actually real and not hypothetical situations on which to legislate" (Fuentes, 2021, Page).

These examples reveal the speculative justifications that gave rise to constitutional reform and that have served as inspiration in other latitudes to begin a debate on a subject that has several layers of complexity. This work, however, will not address the Chilean legislative process on the subject, about which I have already expressed my ideas in other works, so the reference is only contextual.

Thirdly, it is necessary to point out that the rules on neurotechnology's (invasive and non-invasive), as they were proposed in Chile and as they are being promoted in different international forums, do not consider three basic issues: it is a regulation that simplifies what the brain is and its relationship with the body—recent studies show that motivations and emotions depend largely on the microorganisms that populate its digestive system—providing protection to a part of the body and not to the individual as a whole (Cosmelli, 2020), until now neurotechnology's have proven to be safe (Adams et al., 2023; Ruiz et al, 2021, 439; Cornejo & Guiñazú 2023, 111) and in the vast majority of the cases they are used by people who are patients of the health systems to whom solutions based on these technologies are provided.

This point is essential for this essay because it allows us to highlight the fact that most people who resort to this type of advances do so because they need it to recover their health or some capacity that has disappeared or has diminished in such a way that it does not allow them to lead a life in normal or autonomous conditions.

This situation can obviously imply that these people are not qualified to give prior or complete consent because the situation that makes it necessary to use neurotechnology incapacitates the person before its use, or because the person is not able to fully and completely understand the effects or implications that neurotechnology can have on their bodies.

Furthermore, in the case of people with a psychiatric illness, they could even give full and complete consent in appearance but be unaware of what the use of these technologies represents.

Thus, the objective of this work is to reflect on the regulatory model that is being promoted and that is built on consent as a basis of legality for a person to use neurotechnology's. Indeed, Article 3 of the Chilean bill about neurotechnology states that:

Any interference or form of intervention in neural connections or intrusion at the brain level using neurotechnology, brain-computer interface or any other system or device is prohibited, which does not have the free, express and informed consent of the person or user of the device, including in medical circumstances. Even when neurotechnology has the capacity to intervene in the absence of the person's own consciousness in the case of those areas of clinical medical research, the consent determined by the health code in its Title V on clinical trials and the respective regulations will be necessary.

In turn, Article 5 states that:

Any form requesting consent for the intervention, invasive or not, of neurotechnology's, brain-computer interface or other device, must indicate the possible physical effects of its application, the eventual cognitive and emotional effects thereof, the rights and duties, rules on privacy and protection of information, security measures adopted, and contraindications.

The problem is that both rules are put in the scenario that the users of neurotechnology are in full use of their physical and mental faculties and in a position to give consent, and here is one of the many premise errors that this regulatory proposal has, and that is that the majority of people who access these therapies (which are therapies that are not widely available and are expensive) are not in a position to give consent as requested by the rule, that is, informed consent, which in practical terms would result in not being able to give consent and not being able to access the possibility of therapy offered by neurotechnology.

2 The Problem of Consent

Despite the enthusiasm of some authors who advocate the creation of new fundamental rights such as cognitive freedom (Ienca y Andorno, 2017a) or mental privacy (Ienca 2017b), the literature in Chile has documented a series of negative effects of this proposal, in addition to scientific and philosophical inaccuracies, where there are inexact equivalences and where a reductionist idea of cognitive neuroscience is proposed, implying a questionable separation between brain and body (Ruiz et al., 2021, 440).

From a legal point of view, it was already clear in the Chilean academy that the technique of the legislation being promoted would be redundant, since the integrity of the human being (as a whole) is protected by the rights to privacy and physical and mental integrity, which are part of most Western legislation (Zuñiga-Fajuri et al., 2021). The same was done at the international forum, criticizing the inflation of human rights instead of promoting the development of existing ones that should adapt even more to changing social circumstances and technological possibilities (Bublitz, 2022, 13).

On this occasion, this work adds a new objection to this regulatory proposal that is given by the idea of the manifestation of consent and the idea that once expressed, this would become an enabler condition for the use of these technological advances.

In Chilean law, as in much of comparative law, the manifestation of will must be expressed through a declaration that allows it to be known and that, as a rule, must coincide with what the other party proposes. The regulatory proposal aligns with this dogma when it recognizes that "consent must be explicitly formulated by the person who will use the neurotechnology".

In Chile and in comparative law, the intention or will that be not expressed does not have the virtue of producing the agreement that is sought. The rules can be essentially more extreme, since consent must often be specific and determined or practical, which means that what is expected is that, if there is an agreement, the effects resulting from it will be produced.

In this sense, the clarity of the expressed will is a burden that falls on the person who declares consent. There are exceptional cases where a person's silence can constitute a manifestation of will, such as in certain operations where rectifications or clarifications can be made, but they are not. There is the concept of circumstantial silence, which is that accompanied by external facts that allow silence to be attributed, unequivocally, the value of a manifestation of will.

All these characteristics of consent, as an enabling act, emanate from the idea that people are free, autonomous to decide about themselves and their assets. However, this legal construction fails when it is transferred to the field of health and to neurotechnology's.

3 Consent of People Without Physical or Mental Capacity to Do so

One of the many problematic aspects of the regulation on neuro rights is that it is so generic that it does not allow to distinguish between the hypotheses of therapeutic use and of the "commercial" nature. To date, the only applications available are those for medical use, with no record of the existence or occurrence of uses outside that margin that involve healthy people with the effects that the regulation imagines.

In Chile, Law No. 20,584, which establishes the duties and rights of the patient, states in its art. 10 that:

Every person has the right to be *informed*, in a timely and *understandable* manner, by the doctor or other treating professional, about the state of his or her health, the possible diagnosis of his or her illness, the treatment alternatives available for his or her recovery and the risks that this may represent, as well as the expected prognosis, and the foreseeable postoperative process when appropriate, according to his or her age and personal and emotional condition (…) When the condition of the person, in the opinion of his or her treating physician, *does not allow him or her to receive the information directly or suffers from difficulties in understanding or is suffering from an alteration of consciousness, the information referred to in the previous paragraph will be given to his or her legal representative, or failing that, to the person under whose care he or she is.*

Art. 14° also states that:

Every person has the right to grant or deny their will to undergo any procedure or treatment related to their health care (…) *This right must be exercised freely, voluntarily, expressly and in an informed manner, for which it will be necessary for the treating professional to provide adequate, sufficient and understandable information.*

This same article was recently amended in connection with Law No. 21,541, which authorizes health providers to provide care through telemedicine, adding that:

The informed consent of the patient to receive digital health services may be given verbally, in which case the respective institutional and individual provider must record the acceptance or rejection of health care through a written statement in paper format or signed through an electronic system that guarantees its authenticity (…), leaving a record in the clinical record of the safeguards adopted to ensure the person's right to information.

This same law includes hypotheses where this consent must be given by parents or guardians in the case of minors and exceptions to this manifestation of will in art. 15° pointing out:

When the person is unable to express his or her will and it is not possible to obtain it from his or her legal representative, because he or she does not exist or cannot be found. In these cases, *appropriate measures will be adopted in order to guarantee the protection of life.*

However, the same law, regarding biomedical research, states in its art. 28° that "*biomedical research may not be carried out on adults who are not physically or mentally capable of expressing their consent*". The same law refers to advance consent in the case of "People with neurodegenerative or psychiatric illness may give their *informed consent in advance* to be subjects of trials in future research, when they are not in a position to consent or express preference". Indeed:

There are a group of particular issues, such as certain dilemmas of consent in children and incompetent adults, considering that the development and use of these technologies is aimed at people whose autonomy to make decisions is compromised7 and whose participation should be encouraged (Muñoz-Cordal & Junyent-Ruiz 2024, 190–191).

This harmful effect of the regulation promoted by neuro-rights increases the deterioration in all areas of research in Chile. Thus, consent in the Chilean medical field is what has inspired the regulation on neuro rights, but this detailed description has a variety of difficulties that need to be explained. This deterioration coincides with the enshrinement of the rules on consent in the law on the rights and duties of the patient, which is evident from 2012 onwards, where it is possible to show that all clinical studies on neuropsychiatric diseases such as bipolar disorder, schizophrenia, Parkinson's disease, Alzheimer's, multiple sclerosis, etc. fall from that date (Ruiz ct al., 2021, 442–443).

The seriousness of this situation lies in the fact that Chile is the Latin American country with the greatest state resources available for research, which, however, cannot support this kind of research because regulation obstructs them despite being a priority of national and global public policy given the aging of the population. In that sense, the regulation on neuro-rights that is being discussing in the Chilean Parliament and that is intended to be exported to the world, which reproduces the same rule of obtaining consent, would end up putting a stop to research in these areas.

The use of neurotechnology's is the only resource that people with disorders of consciousness have and a regulation such as that formulated by the promoters of neuro rights only deteriorates their position in accessing these improvements that seek a restitution or rehabilitation of the capacities of those who suffer from them, leaving people with some type of disability on the sidelines of well-being.

With all these considerations, the regulation of neuro rights, which, as we postulate, is based on the hypothesis of healthy people who can consent, can generate a problem in all those jurisdictions interested in its incorporation by reducing the possibilities of the most vulnerable people in the health system and who are those people affected by pathologies that compromise their psychiatric or neurological activity, destroying decades of work in the field of promoting human rights in favor of people who suffer from some type of disability (Fins, 2022).

In this sense, the regulation of neuro rights, which, as we postulate, is based on the hypothesis of healthy people who can consent, can generate a problem in all those jurisdictions interested in its incorporation by reducing the possibilities of the most vulnerable people in the health system and who are those people affected by pathologies that compromise their psychiatric or neurological activity, destroying decades of work in the field of promoting human rights in favor of people who suffer from some type of disability, because this legislation overlaps previous rules, which are specific and which in themselves have already contributed to making the panorama stormy for the development of research.

Thus, when the promoters of this right point out that its core content is the promotion of cognitive freedom, mental indemnity, or mental privacy, they overlook the fact that these proto-mental rights completely rule out the positive effects of other fundamental rights, such as the right of access to health, non-discrimination, and equitable access to the progress of science.

Thus, not only is this a redundant regulation because the rights to privacy, freedom and physical and mental integrity are widely recognized in global legal frameworks, but it also tramples on other equally critical rights, the existence of which determines perhaps the most basic of all rights, the right to life.

The European Parliamentary Research Service took the same line when it pointed out in its recent report that:

Our initial analyses have shown that the rights enshrined in the European Convention on Human Rights and the Charter of Fundamental Rights of the European Union can provide an adequate legal framework. However, if new human and/or fundamental rights are introduced, existing rights could be restricted and new regulatory gaps could arise (European Parliament, 2024, 5).

The same report highlighted that creating these new human rights produces a kind of division of human rights and these new, small and specific human rights "run the risk of creating gaps".

This phenomenon had already been warned by the Chilean legal academy when, regarding these eventual risks, it warned about the "normative inflation, which could ultimately lead to the erosion of existing rights, linked to the phenomenon it seeks to address" (Zaror Miralles et al. 2021, 3–4).

4 Conclusion

When a country is an early adopter of a regulation, it always runs risks, however, the way to reduce them to a minimum continues to be the support of evidence, which in the case of study seems completely absent. What is more, the case of the regulation of neuro rights seems dangerously a regulatory scenario with a total lack of evidence of the context where it was approved.

Technology easily makes certain powerful actors, such as parliamentarians and the media, fall into the use and repetition of certain concepts that allow them to give the appearance of a supposed knowledge and up-to-dateness of technology and in that sense the regulation of neuro rights was a successful cultural installation that permeated the press, politicians, some academics and parliamentarians.

The approval of this new constitutional right has not brought any change in the sphere of people's lives because no invasion of mental privacy or cognitive freedom has been verified that has been claimed by invoking this constitutional guarantee. On the contrary, its incorporation into the Constitution and the risk that the bill that is in the Chilean Parliament reactivates its discussion, brings with it the risk of crystallization of a hypothesis of consent that only those who are mentally in full capacity to consent are able to give, leaving aside those who would truly benefit from certain interventions.

The consent structure used by Chilean law is a structure based on a negotiating consent, where there is an economic or exchange logic. It is this idea of consent that was considered in the law on the rights and duties of the patient, and it was this model that the legislators followed to legislate on neurorights. The adoption of this normative structure generates problems because neuroethology, seen as a therapeutic tool, will be used in many cases by people who are not or will not be able to provide prior and informed consent.

Indeed, the central idea on which the consent hypothesis is built is aimed at healthy people or representatives who are in a position to consent, but in the Chilean case the number of obstacles to verifying consent have created a panorama that has been adverse to the promotion of research into psychiatric and neurodegenerative diseases, particularly because these are people who do not have the capacity to do so and who are the ones who directly benefit from biomedical research.

In this sense, as has been recognized by a significant number of experts from the law and medicine and technical panels of the European Union, existing rights amply fulfill their mission of protecting individuals from advances in technology and, on the contrary, the approval of this right, in principle redundant, has at the same time the potential to undermine other fundamental rights in the understanding that it may weaken access to the right to health, to equal opportunities and non-discrimination and to equitable access to the progress of science.

References

Adams, J., Hasan, M., & Thorp, J. (2023). Using neurotechnology in the emergency and safety management for creating a safer work environment. *Journal of Emergency Management, 21*(2), 133–139. https://doi.org/10.5055/jem.0768

Amado, J. A. G (1997). Escritos sobre filosofía del derecho. Ediciones Rosaristas, Colegio Mayor de Nuestra Señora del Rosario.

Atienza, M. (1989). Contribución para una teoría de la legislación. Doxa: Cuadernos de Filosofía del Derecho, (6), 385-404.

Blackstone, S. W. ([1765–1769] 1979). *Commentaries on the Laws of England* (Tomo I) (p. 496). Chicago: Chicago University Press.

Boletín, N. 13.828–19. *Sobre protección de los neuroderechos y la integridad mental, y el desarrollo de la investigación y las neurotecnologías.* Senado. https://tramitacion.senado.cl/appsenado/templates/tramitacion/

Bublitz, J. C. (2022). Novel neurorights: From nonsense to substance. *Neuroethics, 15*, 7. https://doi.org/10.1007/s12152-022-09481-3

CIOMS. (2024). *Real-world data and real-world evidence in regulatory decision making. Working Group report.* Geneva, Switzerland: Council for International Organizations of Medical Sciences. https://doi.org/10.56759/kfxh6213. Licence: CC BY-NC-SA 4.0

Cornejo, Isabel y Guiñazú, María Flavia. (2023). Límites éticos y legales (Neuroderechos) a la utilización de las interfaces cerebro computadora (BCIS) para aumentar las capacidades cognitivas. En: Introducción a la ética y el derecho de la inteligencia artificial (p. 270). Editorial la ley.

Cosmelli, D. (2020). *Reflexiones de un cerebro en un tarro.* La Segunda. https://www.litoralpress.cl/sitio/Prensa_Texto.cshtml?LPKey=4WSHPHAGNTMN2YCN7ERWLA7XAKI5DYDYN3C4ITRZIRNW2QIP4SQA

Secretaría de Estado de Digitalización e Inteligencia Artificial. (2020). *Carta de Derechos Digitales.* España.

Fins, J. J. (2022). The Unintended Consequences of Chile's Neurorights Constitutional Reform: Moving Beyond Negative Rights to Capabilities. *Neuroethics, 15*(3), 1–11.

Fuentes, R. (2021). *El estado de la neurotecnología.* La Segunda. Santiago Chile. https://medicina.uchile.cl/noticias/172289/el-estado-de-la-neurotecnologia

Ienca, M., Andorno, R. (2017a). Towards new human rights in the age of neuroscience and neurotechnology. *Life Sciences, Society and Policy, 13*(1):5. https://doi.org/10.1186/s40504-017-0050-1. Epub 2017 Apr 26. PMID: 28444626; PMCID: PMC5447561. 28 pp.

Ienca, M. (2017b). Preserving the right to cognitive liberty. *Scientific American, 317*(2), 10–10. https://cacmb4.acm.org/opinion/articles/220099-preserving-the-right-to-cognitive-liberty/fulltext

Müller, O., Rotter, S. (2017, Dec 13). Neurotechnology: Current developments and ethical issues. *Front Syst Neurosci, 11*, 93. https://doi.org/10.3389/fnsys.2017.00093. PMID: 29326561; PMCID: PMC5733340.

Muñoz Cordal y Junyent Ruiz, (2024). ¿Qué nos debería preocupar? Algunas ideas sobre los denominados neuroderechos en Chile. *Revista Ius et Praxis, 30*(3), 183–210. https://doi.org/10.4067/S0718-00122024000300183

European Parliament. (2024). *The protection of mental privacy in the areas of neuroscience. Societal, legal and ethical challenges.* Study of the Panel for the future of Science and Technology (p. 117). EPRS.

Ruiz, S., Ramos Vergara, P., Concha, R., Altermatt, F., von Bernhardi, R., Cuello, M., Godoy, J., Valera, L., Araya, P., Conde, E., Toro, P., & Caneo, C. (2021). Efectos negativos en la investigación y el quehacer médico en Chile de la Ley 20.584 y la Ley de Neuroderechos en discusión: la urgente necesidad de aprender de nuestros errores. *Revista Médica De Chile, 149*(3). Recuperado a partir de https://www.revistamedicadechile.cl/index.php/rmedica/article/view/9120

Stieglitz, T. (2021). Why Neurotechnologies? About the purposes, opportunities and limitations of neurotechnologies in clinical applications. *Neuroethics, 14*, 5–16. https://doi.org/10.1007/s12152-019-09406-7

BrainMind Summit. (2020, April 19). *Jack Gallant—brain mapping, brain decoding, and future neurotechnology* [Video]. www.Youtube.com. https://adclicker.io/f/1spa8h

Yuste, R., Goering, S., Arcas, B., et al. (2017). Four ethical priorities for neurotechnologies and AI. *Nature, 551*, 159–163. https://doi.org/10.1038/551159a

Zaror Miralles et al, (2021). Acerca de la necesidad de proteger constitucionalmente la actividad e información cerebral frente al avance de las neurotecnologías: Análisis crítico de la reforma constitucional introducida por la Ley 21.383. *Revista Chilena De Derecho Y Tecnología, 10*(2), 1–10. https://doi.org/10.5354/0719-2584.2021.65650

Zuñiga-Fajuri, A., Villavicencio-Miranda, L., & Zaror-Miralles, D., et al. (2021). Neurorights in Chile: Between neuroscience and legal science. In M. Hevia, *Regulating neuroscience: Transnational legal challenges* (pp. 165–179). Elsevier. https://doi.org/10.1016/bs.dnb.2021.06.001

Principles of Research with Neurotechnologies and Artificial Intelligence: A Comparative Analysis of Latin American and European Laws

Isabel Cornejo-Plaza and Roberto Cippitani

Abstract The chapter examines the ethical and legal implications of neurotechnologies and AI, particularly in research activities where such technologies are increasingly being used. The use of neurotechnologies is regulated in Chile, and other Latin American countries are adopting specific regulations. Within the European Union, there is no legislation on neurotechnologies, but other legal disciplines, such as the Artificial Intelligence Act and the General Data Protection Regulation, are applicable to some extent. In any case, both Latin American and EU legal systems do not focus their attention on the use of neurotechnologies and AI in research activities. Therefore, a key argument is the need for a principles-based ethical framework to guide research in this area. These principles are internal to the legal systems of the European Union and Latin America. The chapter identifies some applicable ethical principles such as human dignity, respect for fundamental rights (including neurorights and cognitive freedom), self-determination, solidarity, precaution and proportionality. It also recalls specific ethical principles applicable to AI and neurotechnologies: human agency and control, transparency, technical robustness and safety, diversity/non-discrimination/fairness, and consideration of cross-cultural perspectives. In summary, the work argues for a strong ethical foundation, based on key principles and values, to guide neurotechnology and AI research and ensure the protection of human rights and dignity as rapid advances in these fields continue.

I. Cornejo-Plaza (✉)
Neurometa; Research Group on Neuro-Rights, Artificial Intelligence and Metaverse; DDLAB (Innovation Laboratory, Technologies and Frontiers of Legal Sciences), Universidad Autónoma de Chile, Santiago, Chile
e-mail: isabelcornejo@derechocivil.cl

R. Cippitani
Chair 'ISAAC' (Individual Rights in Scientific Research) at Universidad Nacional de Educación a Distancia, Madrid, Spain

INDEPAC - Instituto Nacional de Estudios Superiores en Derecho Penal, Ciudad de México, México

Consiglio Nazionale Delle Ricerche, CNR-IFAC, Firenze, Italy

P. López-Silva (ed.), *Contextualizing Neuroprotection: Latin American Perspectives on the Impact of Neurotechnological Development in Life and Society*, Ethics of Science and Technology Assessment 54, https://doi.org/10.1007/978-3-031-96055-0_14

Keywords Neurotechnology · Artificial intelligence · Protection of personal data · Neurodata European Union law · Latin American legislations · Ethics principles

1 Neurotechnologies and Ethical Issues in Research

Neurotechnology was defined in 2007 in the journal Nature Biotechnology as 'any development that can monitor or modify brain function'. Later, in 2019, the OECD echoed this definition, identifying as neurotechnologies all 'devices and processes used to access, study, assess, evaluate, manipulate and emulate the structure and function of neural systems'.

Neuroethics undoubtedly occupies a prominent position among the branches of applied ethics, particularly considering the market demands of artificial intelligence technology companies, which require behavioural research and ethical considerations applied to algorithmic decision-making. Over the last two decades, there has been significant development in neuroethics, largely due to the support of funding agencies and policy makers who have facilitated transnational research and the establishment of scientific networks and platforms. However, "engagement with neuroethics remains limited within traditional academic institutions, including universities and private research organisations, especially in Europe" (Jox & Müller, 2017, Page).

The therapeutic benefits of the use of neurotechnologies are also accompanied by risks and uses in commercial fields, which have aroused the interest of multilateral bodies to establish appropriate governance, including other disciplines such as personal data, notably the General Data Protection Regulation (hereinafter GDPR) and the Artificial Intelligence Act (hereinafter AIA) in the European Union.

The risk of using neurotechnologies without adequate governance based on ethical and legal principles has concerned various multilateral organisations such as the OECD, UNIDROIT, UN, PARLATINO, OEA and, more recently, UNESCO, which is currently preparing ethical regulations based on principles and values.[1] Governance is provided either by ethical guidelines or by objective laws; in the latter case, the most solid experience of protection is the Chilean constitutional reform that, in 2021, modified article 19 of its fundamental charter, which establishes the protection of brain activity.

The neuroethical approach to the study of the brain and central nervous system aims to improve the quality of life of patients (Cornejo-Plaza & Guiñazú 2023). Advances in neuroscience have led to better outcomes for patients with various diseases. For example, studies have shown that deep brain stimulation can improve motor symptoms in people with Parkinson's disease and facilitate their daily activities.

---

[1] UNESCO General Conference at its 42nd session (Resolution 42C/29), establishes an Ad Hoc Expert Group (AHEG) for the preparation of a draft text of Recommendation on the Ethics of Neurotechnology in April 2024.

Neurotechnological interventions such as DBS (deep brain stimulation) or tDCS (transcranial direct current stimulation) have the power to modify individual behaviour, mood or personality.

Neurotechnologies, because of the above considerations, admit two classifications: therapeutic and non-therapeutic and invasive and non-invasive.

The first classification implies the use of neurotechnologies in medical contexts, regulated by biomedical law. On the other hand, non-therapeutic neurotechnologies are those used for commercial or recreational purposes.

As described above, the development of neurotechnologies poses challenges, not only in terms of unauthorised interventions in the brain, but also in the context of neurotechnological procedures deliberately chosen for purposes that go beyond the purely therapeutic and enter the frontier of cognitive enhancement or transhumanism (Cornejo-Plaza & Guiñazú 2023).

A second classification distinguishes between invasive and non-invasive neurotechnologies. The former are technical tools that measure and analyse chemical, physical, acoustic, electrical, magnetic and/or mechanical signals associated with the structure of the nervous system and the functional signals it emits. They can be used to identify, record and/or control features of nervous system activity, understand nervous system function, diagnose pathology, or control external devices (machine-neuron interfaces are often referred to as brain-computer interfaces).

Invasive neurotechnologies include technologies that require brain-computer interfaces to be surgically implanted in or near the brain. Until recently, the invasive surgery required for this subfield posed many risks to humans. Today, the use of nanotechnologies makes it possible to reach the brain via the cardiovascular system, greatly reducing the risks. Non-invasive neurotechnologies are those that are part of brain-computer interfaces that are implemented outside the body (e.g. glasses, visors or headbands) and do not require any kind of surgical implantation.[2]

2 Legislation on Neurotechnologies and Artificial Intelligence

American countries and the European Union are among the most committed to regulating the use of technologies such as neurotechnologies.

However, the regulatory approach may differ in the two continents.

The EU has no specific discipline on neurotechnologies or personal data arising from the use of such technologies, although the European Parliament, in a resolution of 2022,[3] notes its concern about the lack of legislation on neural data (paragraph 36) and calls on the Commission to present a legislative initiative on neuro-rights to

[2] UNESCO (2024).

[3] Resolución del Parlamento Europeo, de 3 de mayo de 2022, sobre la inteligencia artificial en la era digital, (2022/C 465/06).

protect 'the right to identity, free will, mental privacy, equal access to advances in cognitive augmentation and protection against algorithmic bias' (paragraph 247).

On the contrary, there is more sensitivity in this area in America.

At the regional level, one can mention the Model Law on Neurorights for Latin America and the Caribbean, adopted by the Parlatino in Panama on 19 and 20 May 2023.

In addition, some countries, such as Brazil,[4] Mexico and Argentina,[5] are adopting specific rules.

In the United States, the states that are discussing proposals for legislation on neurotechnologies are mainly Colorado, Minnesota and California.

Most notably, in 2021, the President of Chile signed an amendment to the Constitution that establishes the right to life and physical and mental health. The new provision states that "scientific and technological development must be at the service of the people and be carried out with respect for life and physical and mental health". It also states that "the law shall regulate the requirements, conditions and restrictions for its use on human beings and shall protect, in particular, cerebral activity and the information derived from it".

The Chilean constitutional reform, Law No. 21.383, of which Article 19 No. 1 modifies the chapter on fundamental rights and establishes: "The protection of brain activity and the protection of scientific activity". In addition to the above-mentioned laws, a bill entitled 'Regulation of Neurotechnologies' is under discussion. This bill concerns the protection of neuro-rights and mental integrity and seeks to regulate the advancement of research and development of neurotechnologies listed in Bulletin 13.828-19. This bill establishes that neurodata are sensitive biometric data, without distinguishing whether they have been subject to anonymisation or not, as the GDPR has done (Cornejo-Plaza et al., 2024). This article complements the ethical standards for research in neurosciences, neurotechnologies and artificial intelligence, Law No. 20.120 "On scientific research on human beings, their genome and the prohibition of human cloning" of 2006 (Cornejo-Plaza & Saracini, 2023).

Within EU law, in the absence of a specific discipline, the field of neurotechnologies may be regulated by normative acts on specific aspects, such as the discipline on the protection of personal data, such as the European Union Regulation No. 2016/679 (GDPR); consumer protection and the regulation of medical devices; the AIA (i.e. Regulation (EU) 2024/1689 of the European Parliament and of the Council of 13 June 2024 laying down harmonised rules on artificial intelligence).

[4] Two bills have been introduced that seek to protect mental privacy and neuro rights: Bill 29/2023 sought to amend the Brazilian Constitution to include protections for mental integrity and algorithmic transparency. The proposal was presented to the Senate in June 2023 and is pending until a Rapporteur is appointed to review the bill. Bill 522/2022 seeks to amend Brazil's General Data Protection Law (LGPD) to regulate neurodata as a category of sensitive data.

[5] A Neurorights Bill has been presented in the Province of Misiones (2023) Bill 0339-D-2022.

3 Applicability of Legislation Concerning Neurotechnologies and AI to the Research Activities

However, the application of the above mentioned legal sources is not entirely satisfactory.

With regard to the GDPR and the other disciplines protecting personal data, it is not always easy to consider neurodata as 'personal data', although this is the opinion of the Organisation for Economic Cooperation and Development (OECD), which argues that data derived from neurotechnologies are personal data because they are "data relating to the functioning or structure of the human brain of an identified or identifiable individual, containing unique information about his or her physiology, health or mental state".[6]

In fact, the aforementioned regulations apply when the data are personal and, therefore, when they allow the identification of a specific natural person (see Article 2 (f) of Chilean Law 19,628; Article 4 (1) GDPR).

Given the broad interpretation of personal data by the Court of Justice of the European Union, this includes any information 'about' the person concerned.[7]

In addition, Article 4(1) GDPR considers identifiability at the time of processing rather than at the time of collection. Thus, information that would not be identifiable to specific individuals today could become so as a result of technological developments. An analogous case is that of human biological material, which is stored in collections that were created, for example, for medical diagnostic purposes, at a time when there was no way of identifying to whom a tissue sample could belong (other than by associating a first name and surname). Today, this identification is possible and therefore this material contains personal data, but it was collected without complying with the current requirements (informed consent, among others).

The situation is complicated by the fact that neurodata may have a dynamic information content (being an evolving technology), in the sense that it is not easy to distinguish today how it will be possible to read the same set of data in the coming years and which will be disaggregated.

On the other hand, the other EU disciplines do not specifically address neurotechnologies or neurodata, such as the AIA.

In any case, the topic of the research or is subject to exemptions or specific rules (see GDPR) or they are excluded by the legal discipline (such as the AIA).

According to the AIA, research activities fall outside the discipline provided for in Regulation (EU) 2024/1689 (see 'recital' 25).

On the other hand, the GDPR establishes that when 'personal data are processed for scientific, historical research or statistical purposes', European and national legislation may provide for derogations from the rights normally belonging to the data subjects.

[6] OECD, Recommendation of the Council on Responsible Innovation in Neurotechnology (2020), https://legalinstruments.oecd.org/en/instruments/OECD-LEGAL-0457

[7] Court of Justice, judg. December 20, 2017, C-434/16, Peter Nowak v. Data Protection Commissioner, ECLI:EU:C:2017:994, paragraph 34.

As reiterated in the 'recital' 159 of the GDPR:

In order to take into account the specificities of the processing of personal data for scientific research purposes, specific conditions should apply, in particular with regard to the publication or other disclosure of personal data in the context of scientific research.

Therefore, specific rules on consent could be adopted ('recital' no. 33 GDPR), further processing (Article 5(1)(b) GDPR) and storage of data (Article 5(1)(e) GDPR). Also in this case, scientific purposes allow for derogations from the rules on data processing in general. The exceptions provided for research purposes are only allowed if "appropriate safeguards ... for the rights and freedoms of the data subject" are put in place. These safeguards shall ensure that technical and organisational measures are taken, in particular to ensure compliance with the principle of data minimisation (see Article 89(1) GDPR).

Similarly, in the case of Latin American legislation, and in particular the Chilean law, although the latter is more focused on the issue of neurotechnology, it does not take into account scientific research. As a matter of fact, a judgement of the Chilean Supreme Court (ruling rol N 1.080-2020 issued on August 9, 2023, by the Supreme Court of Justice of Chile, Girardi v. Emotiv Inc.), which is currently the only ruling in this area, shows, neurotechnologies give rise to numerous ethical and legal problems that cannot be limited to the protection of privacy (Cornejo-Plaza et al., 2024; see also Andorno, 2023).

The Chilean Supreme Court ruling raises other important questions.

As underlined in the Girardi v. Emotive Inc. judgment, "the development of new technologies involves more and more aspects of the human person, aspects that were unthinkable a few years ago", and the State is expected to act "to prevent and anticipate their possible effects, in addition to directly protecting human integrity in its entirety, including privacy and confidentiality" (see paragraph 8 of the judgment). Thus, the judgment points out that privacy is an (important) aspect of human integrity or, in other words, of dignity and other human rights such as cognitive freedom, freedom of thought and identity (no 2023b; Ligthart et al., 2023; Lavazza & Giorgi, 2023; Farahany, 2023).

It is particularly relevant that, in the seventh section of the ruling, the Chilean Supreme Court emphasises the importance of obtaining informed and specific consent from individuals involved in scientific activities, citing both international sources and national legislation on scientific research (Law 20.120), regardless of the anonymisation of the data.

4 Regulation for Principles

Due to the fact the scientific research is not covered by legislation on AI or neurotechnologies, the above mentioned judgment of the Supreme Court of Justice of Chile, the EU legislation and other sources, such as the documents issued by International

Organisations (e.g. UNESCO) refer to the necessity to use the 'ethics principles', especially in research activities.

In this sense, it is stated in European sources that AI must be *ethical*, ensuring respect for ethical principles and values; to make reliable AI a reality it must comply with the law and must also be consistent with ethical standards.[8]

The AIA provides that the use of AI must comply with EU and International legal sources and the Ethics guidelines for trustworthy AI of the High-Level Expert Group on Artificial Intelligence (see the 'recitals' 7, 8, 27 AIA).[9] In particular 'any research and development activity should be carried out in accordance with recognised ethical and professional standards for scientific research and should be conducted in accordance with applicable Union law' ('recital' 25 AIA).

On the other hand, the respect of the ethics principles in research activity is provided by GDPR (see 'recital' no. 33) and other legal texts concerning science and technology.

The ethical principles can be considered as legal concepts that aim to balance techno-science and the rights or interests that may be affected by an inappropriate use of technologies.

In juridical terms, the 'principles' are expressed by several provisions within the legal system (see, for example, Betti 1971, 312). They ensure the coherence of the entire legal system, and therefore they constitute an important guide in the interpretation and application of norms.

They can be identified in many legal fields but are particularly useful to regulate situation associated to science and technology, subject to the evolution of the technoscience and to the necessity to respect the freedom of research, recognised at national, international and supranational levels (see for example Article 13 of the Charter of Fundamental Rights of European Union) (about academic freedom see Cornejo-Plaza & Cippitani 2023a; Cippitani, 2015).

Instead of establishing overly rigid legislative rules that are difficult to change except through a lengthy process, it is preferred to lay down framework rules and then refer to more flexible principles that can be specified by the scientific community itself or by interpreters.[10]

For these reasons, in the field of ethics so-called 'principlism' is very popular; this is the use of principles to solve ethical issues, particularly when there is a conflict between different interests (European Commission 2010, 28) (Cippitani 2023a).

Within the European Union's legislative system and in international legal sources (see in particular Draft of the Ethics of Neurotechnology of UNESCO, hereinafter

[8] European Commission: Directorate-General for Communications Networks, Content and Technology and Grupa ekspertów wysokiego szczebla ds. sztucznej inteligencji, Ethics guidelines for trustworthy AI, Publications Office, 2019. https://data.europa.eu/doi/10.2759/346720

[9] See also the European Parliament resolution of 20 October 2020 with recommendations to the Commission on a framework of ethical aspects of artificial intelligence, robotics and related technologies, 2020/2012(INL).

[10] Some authors (Borbón Rodríguez et al., 2020; Fins 2022) argue that in many case the present legislative approach to the neurotechnology can be reduce the scientific research in that field. Different opinion is expressed by Cornejo-Plaza, 2023.

referred as ‘Ethics of Neurotechnology’) one can identify some ethical principles applicable to the elaboration and the use of AI and neurotechnology.

Adopting the approach of the document Ethics of Neurotechnology, the following paragraphs will consider general principles (called ‘values’) and specific principles focused on the above-mentioned technologies.

5 Values Applicable to AI and Neurotechnologies.

Within the EU law and the international texts, such as the document Ethics of Neurotechnology it is provided the use of the technologies at issue has to comply with some fundamental value (Cippitani, 2023b).

(a) Dignity

Dignity is the cornerstone of the international sources for the protection of human rights (see the preamble of ‘International Covenant on Civil and Political Rights’ adopted by UN on 1966) and in national and supranational legislation (see Article 2 of the Treaty on European Union).

It can be defined as the inviolable and non-negotiable core of every human right, arising from the 'primacy' of the individual in modern legal systems. It is seen as a 'conceptual matrix' that protects humanity and vulnerable people (Mislawski, 2010).

In particular, dignity must be respected in research (see Article 3 of the EU Charter of Fundamental Rights) and in other projects and activities involving individuals (see Andorno, 2009; see also the preambles of both the Charter of Fundamental Rights and the Oviedo Convention), and especially in the field of AI (see, for example, the White Paper on Artificial Intelligence; see also the preamble of the AIA) and neurotechnology (Section IV.1.1 (35), Ethics of Neurotechnology).

Persons involved in such activities must therefore not be subjected to degrading treatment, either physically or psychologically, and must not suffer unnecessary stress or pain. Dignity also implies the prohibition of the economic exploitation of a person and of his body and any part of it (see Charter of Fundamental Rights).

(b) Respect of fundamental rights arising from the technologies

As a consequence of the above, AI and neurotechnology have to respect the fundamental rights of the natural persons (see ‘recital’ 1 et al. AIA; see also Parliament’s resolution on robotics, para. 10).

Such technologies must not violate or endanger the private and family life of individuals, especially the most vulnerable; freedom of expression and information; freedom of assembly and association; consumer and workers’ rights; the right to health; security; the environment.

The document Ethics of Neurotechnology of UNESCO (see paragraphs 36 et sub.) identifies come specific rights, such as the ‘neurorights’ and/or ‘cognitive liberty’ which imply that ‘No human being or human community should be harmed or subordinated, whether physically, economically, socially, politically, culturally or mentally, by the use of neurotechnology', as well as the human health 'as a state of complete physical, mental and social well-being and not merely the absence of disease or infirmity' (paragraph 41).[11]

Under AIA and the document of UNESCO on the Ethics of Neurotechnology is central the protection of the personal data (see in particular Article 2 (7) AIA) and the mental privacy (see Ethics of Neurotechnology, paragraph 75).

In addition, some authors (Marinaro & Muñoz, 2024) propose to apply to the neurorights a special proceeding for constitutional protection which they call '*habeas cogitationem'*

(c) Self-determination

Individuals involved in any kind of activity (commercial, scientific, health, educational etc.) have the right to express their free and informed consent in the application of the principle of self-determination.

Although the EU sources referring to AI do not speak directly about the need for consent, they do so implicitly when they for example state the obligation to respect the fundamental rights of individuals including the protection of personal data.

In any case, the concern that the use of AI should be based on consent is expressed in European documents, as well as in the European Parliament's Report on AI in the educational, cultural and audiovisual sectors[12] .

The document Ethics of Neurotechnology of UNESCO is more clear with respect to the self-determination, considered as expression of the freedom of thought, stating that 'In the interaction of a human being with a neurotechnology application, the protection of freedom of thought of the person should cover both the forum externum, i.e., the expression or manifestation of thought through behaviour, and the forum internum, i.e. the underlying neurobiological, sensory, motor, mental state processing'. This in order to avoid manipulation and protection from undue interference (Ethics of Neurotechnology, paragraph 72). The individuals "have the right to self-determination to make free, informed, and voluntary decisions about whether they want to use a certain neurotechnology application or refuse to do so" (Ethics of Neurotechnology, paragraph 73).

Beyond the discussion of whether neurotechnologies should have protection based on new constructions of human rights or reconfigurations of them, what is certain is that all ethical declarations of neurotechnologies must and continue to be based on respect for human dignity and on the system of human and fundamental rights, as Bublitz (2024) rightly points out:

Existing human rights draw several layers of protection around the person. The basic layer is formed by the rights to bodily and mental integrity as well as the right to privacy which, in conjunction, cover interferences through neurointerventions and

[11] About the concept and taxonomy of 'neuroright' (see Ienca & Andorno 2017; Yuste et al., 2017). More recently see López-Silva & Valera (2022).

[12] European Parliament, Committee on Culture and Education, REPORT on artificial intelligence in the educational, cultural and audiovisual sectors, 19 April 2021—(2020/2017(INI)).

neuroimaging. Notably, this protection is comprehensive, it covers all technological interferences that directly interact with the CNS, the modus operandi of the neurotechnologies of today and tomorrow (Bublitz, 2024; Gilbert & Russo, 2024).

(d) Solidarity

Today constitutional systems are based on the protection of fundamental rights in the context of the duty of solidarity (Cippitani, 2010, 2022).

In its modern meaning (Peces-Barba Martinez et al., 2001) solidarity can be defined as the duty to protect the interests of others, in particular vulnerable persons.

Such a principle is applicable in all social and legal relations, both vertical (between authorities and citizens) and horizontal (between individuals, e.g. family and contractual relations), including with regard to property.

From the point of view of the ethics of science and technology, the principle of solidarity means that respect for people's dignity and fundamental rights, in particular of those who are marginalised or in vulnerable situations, such as children, elderly people, persons with disabilities.

This principle is highlighted by the AIA[13] and by the documents concerning neurotechnology' (Ethics of Neurotechnology paragraph 38).

(e) Prevention and precaution

The precautionary principle is the basis for a risk assessment decision-making process. This process must identify the degree of uncertainty involved in the evaluation of scientific information, and then it is necessary to establish the level of risk 'reasonable' to society which, in any case, cannot affect the dignity and integrity of individuals. The risk must be present (not merely hypothetical) and cannot be ruled out, even if it is not safe.

In the specific case of AI, EU sources take into account that the use of AI may give rise to risks from various points of view which may be linked to cyber threats, personal security (e.g. in relation to new uses of AI, as in the case of home appliances) (See White Paper on AI, para. 5).

According to the above-mentioned Communication 'Building Trust in Human-Centred AI').

On the other side, AIA use a risk-based approach classifying the AI systems implying an unacceptable risk (prohibited by Article 5 AIA) and those of 'high risk'. While the AI systems with unacceptable risks are prohibited, the producers and deployers of high risks systems are subject to several obligations.

(f) Proportionality

Another general principle that also applies to AI and neurotechnology is 'proportionality' whereby technoscience must be carried out by means that are not excessive in relation to the ends to be achieved. The means and ends must be legitimate and must avoid risks that may affect the interests protected by the legal system.

---

[13] European Parliament Resolution of 20 October 2020 with recommendations to the Commission on a framework for the ethical aspects of artificial intelligence, robotics and related technologies (2020/2012(INL)), para. 38.

European documents, for example, state that the use of AI systems to fight crime is admissible only if they are necessary and proportionate[14] and that in the field of education, the application of AI should be proportional, in the sense that it should not be used in an excessive or useless way, cancelling out other education methodologies or even (as is perhaps exaggeratedly claimed) replacing teachers (see Cornejo-Plaza & Cippitani, 2023b).[15]

In the field of neurotechnology, the document of UNESCO on ethics issues affirms that the proportionality implies ensuring: '(a) appropriate and proportional to the objective and expected benefits that is aimed to be achieved; (b) do not infringe upon the foundational values of this document; (c) appropriate to the context and target user group; (d) based on safety principles and rigorous scientific evidence' (Ethics and Neurotechnology, paragraph 80).

6 Specific Ethical Principles

EU legislation and other legal sources on AI and neurotechnologies refer to principles specifically applicable to these technologies (see Cippitani, 2023a, 2023b with respect to AI).

This is the case of the AIA, which recalls the principles elaborated by the High-Level Expert Group on AI (see generally Recital 27 of the AIA) or the documents of international organisations such as UNESCO.

(i) Human agency and control.

This principle means that 'AI systems will be developed and used as a tool that serves humans, respects human dignity and personal autonomy, and operates in a way that can be appropriately controlled and supervised by humans' ('recital' 27 AIA; see also the 2019 Communication 'Generating co-finance...' para. 2.2.i).

This implies the prohibition of automated decisions relating to individuals (see Article 22 GDPR).

Oversight should be provided by governance mechanisms, such as the human-in-the-loop approach, human-on-the-loop supervision or human-in-command.

In the case of AI applied to education, the principle under discussion may be useful in the case of evaluation of educational outcomes by teachers in terms of subjects, degrees, doctorates, exams, etc.

(ii) Transparency.

Transparency 'means that AI systems are developed and used in a way that allows for appropriate traceability and accountability, while making people aware that they

---

[14] See 2020 White Paper on Artificial Intelligence, cit., para. 6.

[15] European Parliament Resolution of 20 October 2020 with recommendations to the Commission on a framework for the ethical aspects of artificial intelligence, robotics and related technologies (2020/2012(INL)), para. 118.

are communicating or interacting with an AI system, as well as properly informing users about the capabilities and limitations of that AI system and affected persons about their rights' (see 'recital' 27 AIA).

The 2019 Communication on 'Building trust in human-centred artificial intelligence' (see paragraph 2.2.iv) foresees that it is important to record and document both the decisions made by the systems and the whole process (including a description of the data collection and tagging, and a description of the algorithm used) that led to the decisions. Measures must be taken to ensure the accountability of the algorithmic decision-making process.

As stated in the Ethics of Neurotechnology document:

Scientific research and development focused on neurotechnology should be transparent, based on the best evidence and consistent with international principles of responsible conduct of research and scientific integrity. This requires, among other things, that clinical trials involving neurotechnology applications comply with guidelines on pre-registration of trials, fair selection of participants, approval by independent ethics committees, and responsible and transparent communication of scientific findings (paragraph 44).

(iii) The term 'technical robustness and safety' means that 'AI systems are developed and used in a way that allows for robustness to problems and resilience to attempts to modify the use or performance of the AI system to enable unlawful use by third parties and to minimise unintended harm' ('recital' 27 AIA).

This principle implies that AI systems should be robust and resilient to adverse events, such as unlawful use by third parties, and minimise unintended harm (Article 15(4) AIA). Systems must operate safely and reliably and avoid any adverse effects, which is necessary from a technical and societal point of view. From a data processing perspective in particular, this principle is linked to the issue of information security and the security of personal data.

(iv) Diversity, non-discrimination and fairness.

The legal sources point out that the use of AI and neurotechnology must be used in a non-discriminatory manner.

As stated in the preamble of the AIA:

Diversity, non-discrimination and fairness mean that AI systems are developed and used in a way that is inclusive of different stakeholders and promotes equal access, gender equality and cultural diversity, while avoiding discriminatory effects and unfair biases prohibited by Union or national law ('recital' 27 AIA).

UNESCO, in its document Ethics and Neurotechnology, highlights the risks that may be used to "reproduce or reinforce existing discriminatory practices or introduce new forms of discrimination based on neurological or neuropsychological characteristics" (paragraph 81). In fact:

Information about a person's mental health or predisposition to develop a neurological or psychiatric disorder should not be used to determine decisions related to employment, health insurance, etc., unless this poses a risk to the person concerned, to other groups or to social security.

These forms of discrimination should not "aggravate or reinforce existing inequalities between individuals, groups or countries". Nor should they create new inequalities based on cognitive or other neurobiological characteristics (paragraph 82).

(v) Consideration of cross-cultural perspectives on human knowledge and its sharing.

The draft UNESCO Recommendation on Ethics and Neurotechnology emphasises the importance of sharing knowledge about the human nervous system in a respectful manner between different cultures and communities. This aspect is very relevant to the mestizo and indigenous populations that make up Latin America. Key points include

- Building trust: Sharing knowledge builds trust and cooperation between different cultural groups.
- Community engagement: Engaging communities and maintaining a two-way exchange strengthens collective well-being and promotes health.
- Diversity in research: Recognising the unique perspectives of diverse groups, especially indigenous and traditional knowledge, enriches scientific research. Research should be conducted with the explicit consent of these groups to prioritise their interests.
- Contextual approaches: Emphasises the need to understand neuroscientific knowledge within specific cultural frameworks and advocates for regulations that respect and incorporate diverse perspectives on neurocognitive diversity.

This framework promotes a more inclusive vision of the health sciences and advocates for collaborations that recognise the cultural significance of knowledge. If you would like more details on a specific point, please let me know: "Contextual and cultural relativism calls for consideration of a possible non-universal understanding and priority of current neuroscientific knowledge, so any regulatory framework must be contextual and culturally aware, responsible, respectful and inclusive of neurocognitive cultural diversity" (Herrera-Ferrá et al., 2022).

(vi) With regard to the principle of accountability, the Communication "Building trust and confidence in human-centred AI" (see paragraph 2.2.vii) states that mechanisms should be put in place to ensure the responsibility and accountability of AI systems and their results, both before and after their implementation, as well as the possibility of internal and external auditing of the use of AI systems. In terms of auditability, it is the ability to evaluate algorithms, data and design processes, which must be assessed by internal and external auditors and the availability of evaluation reports that will contribute to the reliability of the technology.[16]

---

[16] See paragraph. 8.3. of the Proposal for a Regulation of the European Parliament and of the Council on ethical principles for the development, deployment and use of artificial intelligence, robotics and related technologies. In European Parliament resolution of 20 October 2020 with recommendations

According to the document Ethics of Neurotechnology "All stakeholders involved in the development, deployment, and management of neurotechnology should consider accountability as a central value" (paragraph 64). This involves "adherence to regulatory and professional standards, ensuring that all activities respect personal privacy, dignity, and environmental sustainability" (paragraph 65).

7 Conclusions

The convergence of neurotechnologies and artificial intelligence poses global and systemic ethical and legal challenges that require urgent attention and a collaborative approach between researchers, regulators and communities to ensure that research practices are ethical and respect the fundamental rights of individuals. In Latin America, the regulation of neurotechnologies should consider the principles currently being worked on in the UNESCO proposal for a regulation on the Ethics of Neurotechnologies that takes into account the cultural aspects specific to the Latin American region, which has closely followed the EU in terms of regulation, especially with regard to the protection of personal data and, more recently, the AIA. However, specific legislation on neurotechnologies is lacking in the EU, in contrast to the rapid implementation of legislation in Latin America, such as the constitutional reform in Chile that protects brain activity and establishes neurotechnological rights.

Ethical principles must be integrated into all aspects of research and development of neurotechnologies to ensure their responsible, supportive and equitable use.

8 Financial Support

This work was supported by European Union, Executive Agency EACEA: Jean Monnet Module e-RIDE (Ethics and Research Integrity in the Digital Age (2024–2027), Grant Agreement N° 101175756; Jean Monnet Chair Gobernanza y Regulación en la Era Digital (GovReDig), Grant Agreement No. 101127331 and Jean Monnet Artificial Intelligence and European Private Law (2024–2027) Grant Agreement N° 1011243.

References

Andorno, R. (2009). Human dignity and human rights as a common ground for a global bioethics. *In Journal of Medicine and Philosophy., 34*, 223–240.

---

to the Commission on a framework for the ethical aspects of artificial intelligence, robotics and related technologies (2020/2012(INL)). OJ C 404, 6.10.2021, 63–106.

Andorno, R. (2023). *UNESCO Report for the month of October 2023 on"Neurotechnologies and Human Rights in Latin America and the Caribbean: challenges and proposals for Public Policy"*. https://unesdoc.unesco.org/ark:/48223/pf0000387079/PDF/387079spa.pdf.multi

Borbón Rodríguez, D. A., Borbón Rodríguez, L. F., & Laverde Pinzón, J. (2020). Análisis crítico de los NeuroDerechos Humanos al libre albedrío y al acceso equitativo a tecnologías de mejora. *IUS ET SCIENTIA, 6*(2), 135–161. https://doi.org/10.12795/IETSCIENTIA.2020.i02.10

Bublitz, C. (2024). Neurotechnologies and human rights: Restating and reaffirming the multi-layered protection of the person. *The International Journal of Human Rights, 28*(5), 782–807. https://doi.org/10.1080/13642987.2024.2310830

Cippitani, R. (2010). *La solidarietà giuridica tra pubblico e privato.* Perugia: Istituto per gli Studi Economici e Giuridici.

Cippitani, R. (2015, 24–26 June). Academic freedom as a fundamental right. In *1st international conference on higher education advances, HEAd'15, Universitat Politècnica de València* (pp. 552–558). Valencia: Universitat Politècnica de València. ISBN: 978-84-9048-340-4. https://doi.org/10.4995/HEAD15.2015.1522

Cippitani, R. (2022). La solidaridad institucional europea frente a las situaciones de crisis. In *Carlos Molina del Pozo, Virginia Saldaña Ortega, Hacia la construcción de un verdadero proyecto federal para la Unión Europea* (pp. 64–87). Colex, A Coruña. ISBN: 978-84-1359-661-7

Cippitani, R. (2023a). Ethical principles and legal provisions. In V. Colcelli, R. Cippitani, C. Brochhausen-Delius, R. Arnold (Eds.), *GDPR requirements for biobanking activities across Europe.* Springer. https://doi.org/10.1007/978-3-031-42944-6_20

Cippitani, R. (2023b). *La dimensión jurídica del Espacio Europeo de la Investigación.* CNR Editions, Roma. ISBN printed edition: 978-88-8080-590-8, ISBN electronic edition: 978-88-8080-591-5, https://doi.org/10.19270/LawScience2_2023

Cornejo Plaza, M. I. & Cippitani, R. (2023). Consideraciones éticas y jurídicas de la IA en Educación Superior: Desafíos y Perspectivas. In *Revista de Educación y Derecho (Education and Law Review), Núm. 28 (2023), Inteligencia artificial e implicaciones jurídicas.* ISSN printed: 2013-584X, ISSN online: 2386–4885. https://doi.org/10.1344/REYD2023.28.43935

Cornejo-Plaza M. I., & Cippitani R., (2023, Dec 21) La libertad de investigación en el proceso de propuestas de nueva Constitución en Chile. In *Estudios Constitucionales, Núm*(pp. 88–119).Especial. ISSN 0718-0195

Cornejo-Plaza, M. I. & Guiñazú, F. (2023). Límites éticos y legales (neuroderechos), a la utilización de las interfaces cerebro computadora (BCIS) para aumentar las capacidades cognitivas. In M. Azuaje (Ed.) *Introducción a la ética y el derecho de la inteligencia artificial.* La Ley.

Cornejo-Plaza, M. I., Cippitani, R., & Pasquino, V. (2024) Chilean Supreme court ruling on the protection of brain activity: neurorights, personal data protection, and neurodata. *Frontiers in Psychology, 15.* https://doi.org/10.3389/fpsyg.2024.1330439

Cornejo-Plaza, M. I., & Saracini, C. (2023). On pharmacological neuroenhancement as part of the new neurorights' pioneering legislation in Chile: A perspective. *Frontiers in Psychology, 14,* 1177720. https://doi.org/10.3389/fpsyg.2023.1177720

Cornejo-Plaza, M. I. (2023). Chilean neurorights legislation and its relevance for mental health: criticisms and outlook. *Revista Salud Mental, 46*(5), 269–273. https://doi.org/10.17711/10.17711/SM.0185-3325.2023.034

European Commission. (2010). European textbook on ethics in research. Publications Office of the European Union, Luxembourg. https://op.europa.eu/en/publication-detail/-/publication/12567a07-6beb-4998-95cd-8bca103fcf43

Farahany, N. A. (2023). *The battle for your brain: Defending the right to think freely in the age of neurotechnology.* Martin's Press.

Fins, J. J. (2022). The unintended consequences of Chile's neurorights constitutional reform: Moving beyond negative rights to capabilities. *Neuroethics, 15*(26), 1–3. https://doi.org/10.1007/s12152-022-09504-z

Gilbert, F., & Russo, I. (2024). Neurorights: The land of speculative ethics and alarming claims? *AJOB Neuroscience, 15*(2), 113–115. https://doi.org/10.1080/21507740.2024.2328244

Herrera-Ferrá, K., Muñoz, J. M., Nicolini, H., Saruwatari Zavala, G., & Martínez Bullé Goyri, V. M. (2022). Contextual and cultural perspectives on neurorights: Reflections toward an international consensus. *AJOB Neuroscience*. https://doi.org/10.1080/21507740.2022.2048722

Ienca, M., & y Andorno, R. (2017). Towards new human rights in the age of neuroscience and neurotechnology. *Life Sciences, Society and Policy, 13*(1), 1–27. https://doi.org/10.1186/s40504-017-0050-1

Jox, R.J., & Müller, S. (2017). Investigación en neuroética en Europa. In E. Racine, J. Aspler (Eds.) *Debates sobre la neuroética. Avances en Neuroética*. Springer, Cham. https://doi.org/10.1007/978-3-319-54651-3_15

Lavazza, A., & Giorgi, R. (2023). Philosophical foundation of the right to mental integrity in the age of neurotechnologies. *Neuroethics, 16*, 10. https://doi.org/10.1007/s12152-023-09517-2

Ligthart, S., Ienca, M., Meynen, G., Molnar-Gabor, F., Andorno, R., Bublitz, C., Catley, P., Claydon, L., Douglas, T., Farahany, N., Fins, J. J., Goering, S., Haselager, P., Jotterand, F., Lavazza, A., McCay, A., Wajnerman Paz, A., Rainey, S., Ryberg, J., & Kellmeyer, P. (2023). Minding Rights: Mapping Ethical and Legal Foundations of 'Neurorights'. *Cambridge quarterly of healthcare ethics: CQ: the international journal of healthcare ethics committees*, 1–21. Advance online publication. https://doi.org/10.1017/S0963180123000245

López-Silva, P., & Valera, L. (Eds.). (2022). *Protecting the mind: Challenges in law, neuroprotection, and neurorights (Ethics of Science and Technology Assessment Book 49)*. Springer.

Marinaro, J. Á., & Muñoz, J. M. (2024). "You shall have the thought": Habeas cogitationem as a new legal remedy to enforce freedom of thinking and neurorights. *Neuroethics, 17*(1), 1–22.

Mislawski, R. (2010). Dignité, autonomie, vulenrabilité: Aproche juridique. In E. Hirsch (coordinated by). *Traité de bioétique. Vol. I. Fondements, principles, repéres*. Toulouse, Érés

Peces-Barba Martinez, G., et al. (2001). *Historia de los Derechos fundamentales, t. II, Siglo XVII, vol. I, El contexto social y cultural de los derechos. Los rasgos generales de evolución*. Madrid: Dikinson.

UNESCO. (2024). *Draft of the ethics of neurotechnology*. https://unesdoc.unesco.org/ark:/48223/pf0000389768

Yuste, R., Goering, S., Arcas, B., et al. (2017). Four ethical priorities for neurotechnologies and AI. *Nature, 551*, 159–163. https://doi.org/10.1038/551159a

AI and Neurotechnology Regulation in Chile: A Comparative Case Analysis

Gabriela Arriagada-Bruneau, Abel Wajnermann, and Pablo López-Silva

Abstract In 2021, Chile became the first country to discuss two initiatives addressing the potential misuse of neurotechnologies: a constitutional amendment (Bill 13827-19) and a bill to create neurorights (Bill 13828-19). While the constitutional reform was swiftly approved, the bill for the creation of specific neurorights remains under discussion in the Chamber of Deputies. Despite initial support, both initiatives currently face criticism. Meanwhile, Chile has advanced in regulating AI systems through legislative proposals (Bills 16821-19/15869-19). A study by (Humeres et al. in Big Data Society, 12(1), 2025) identifies three key controversies in Chile's AI regulation process: deflection of technology liability, instrumentalization of policy, and moralization of technology use. This chapter explores how this analysis focused on sources of controversy could inform the neurorights debate currently in need of a systematic socio-political analysis. Applying this framework could help categorize the diverse positions within the discussion, providing a more comprehensive understanding of its development.

Keywords Artificial intelligence · Neurotechnology · Neurorights · Regulation

---

G. Arriagada-Bruneau (✉)
Instituto de Éticas Aplicadas (IEA), Instituto de Ingeniería Matemática y Computacional (IMC), Pontificia Universidad Católica de Chile, Santiago, Chile
e-mail: gcarriagada@uc.cl

Centro Nacional de Inteligencia Artificial (CENIA), Santiago, Chile

A. Wajnermann
Instituto de Éticas Aplicadas (IEA), Pontificia Universidad Católica de Chile, Santiago, Chile

P. López-Silva
School of Psychology and Research Centre in Cognition and Language Development, Universidad de Valparaíso, Valparaíso, Chile

Millenium Institute for Research in Depression and Personality (MIDAP-Chile), Valparaíso Institute for Complex Systems, Valparaíso, Chile

Collaborating Centre for Values-Based Practice in Health and Social Care, University of Oxford, Oxford, UK

P. López-Silva (ed.), *Contextualizing Neuroprotection: Latin American Perspectives on the Impact of Neurotechnological Development in Life and Society*, Ethics of Science and Technology Assessment 54, https://doi.org/10.1007/978-3-031-96055-0_15

1 Introduction

In 2021, Chile became the first country in the world to discuss two initiatives aimed at protecting citizens from the potential misuse of medical and commercial neurotechnologies, namely, a bill to amend the first section of Article 19 of the national constitution (Bill 13827-19), and a bill to create specific neurorights with legal force (Bill 13828-19). Following ethical concerns raised in the context of multiple international scientific research projects focused on neurotechnological development, these two initiatives were aimed to show the government's explicit concern about how neurotechnologies could negatively impact society at different levels if not properly regulated. However, despite its initial popularity, both initiatives have received a number of criticisms over the last years (Borbón & Borbón, 2021; Bublitz, 2022; Hertz, 2023; Ligthart et al., 2023; Susser & Cabrera, 2023; Zuñiga-Fajuri et al., 2021; López-Silva & Madrid. 2021, 2022, 2024). Indeed, while the constitutional reform was promptly approved, the bill for the creation of specific neurorights is still under discussion in the Chilean's Chamber of Deputies. In parallel, and contrasting with the tone of the neurorights discussion, over the last two years Chile has been moving forward in the implementation of legal changes aiming at regulating the use and application of Artificial Intelligence (AI) technology systems (Bill proposal 16821-19/15869-19). A detailed examination of the socio-political dynamics of AI regulation in Chile offered by Humeres et al. (2025) identifies three main sources of controversy during the legislative process, namely, the deflection of technology liability, the instrumentalization of technology policy, and the moralization of technology use. In this chapter, we examine how the identification of these sources of controversy could inform the examination of how the discussion about neurorights in Chile has developed. In contrast to the AI case, the analysis of the socio-political dynamics of the neurorights debate in Chile has not been systematized so the application of this analytic framework could allow the elaboration of a comprehensive characterization of different positions that have been defended in this context.

2 AI Regulation in Chile

Traditional technology governance models face significant challenges when applied to the analysis of the ethical issues associated with the use of AI technologies. While such models are often thought to regulate predictable component-based systems, AI technologies are inherently opacity systems characterized by emerging behaviors. As Judge et al. (2024) argue, the unpredictability of AI systems, particularly those developed through machine learning (ML), undermines the conventional assumption that design and operational compliance can ensure safety and reliability. As a consequence, there seems to be an urgent need to develop adaptive and robust regulatory models tailored to AI's complexities and the ecosystem in which they operate.

One major debate in this context centers on whether AI should be governed through sector-specific regulations or horizontal approaches encompassing all applications in society (Finocchiaro, 2024). While the European Union's AI Act exemplifies the latter (offering a framework with risk-based classifications and harmonized rules), other approaches emphasize the tailoring of regulations to specific domains such as healthcare, data protection, and justice (Kane, 2021; Laptev & Feyzrakhmanova, 2024; Palaniappan et al., 2024; Tschider et al., 2024). This dichotomy seems to reflect broader tensions between fostering innovation and ensuring ethical compliance, a balance that is particularly challenging given the rapid pace of technological advancement and global competition (de Bruin, 2024; Staszczyk, 2024). Furthermore, the international nature of AI development underscores the need for transnational regulatory coordination. Zaidan and Ibrahim (2024) highlight the limitations of fragmented national approaches, emphasizing the risks of uneven development and the potential for regulatory arbitrage. The authors advocate for the establishment of a global regulatory authority to harmonize standards and ensure equitable distribution of AI's benefits while mitigating its risks. This perspective aligns with calls for a precautionary yet enabling framework that addresses both innovation and responsibility.

As aforementioned, Humeres et al. (2025)'s study explored how legislative debates regarding AI regulation in Chile are shaped through specific processes that temper potentially contentious discussions. This case study not only highlights the specific processes employed in Chile but, as we will argue, it also offers insights applicable to broader global regulatory efforts that can apply to the field of neurotechnologies, as similar challenges apply to both discussions. In their analysis, the authors analyze the discussions in the Chilean Chamber of Deputies regarding AI regulation, which occurred within the "Future, Science, Technology, Knowledge, and Innovation" Commission, which initiated the deliberation of a draft bill in mid-2023. Participants in these debates included a mix of parliamentarians from various political parties, state bureaucrats, and invited experts primarily from academia. These experts represented fields such as engineering, law, and social sciences. The discussions, spanning 11 sessions, were held in person and were accessible to the public through broadcasts on the National Congress's website. Citizen participation was limited to a brief consultation period to gauge public approval of legislating AI. This diverse yet focused assembly sought to explore the implications of regulating AI systems, robotics, and related technologies across various applications, reflecting a wide range of perspectives on the opportunities and challenges posed by AI. In the remainder of this chapter, we examine how the three main sources of controversy identified by the authors in the AI regulation process in Chile can inform our understating of three main domains in the case of neurorights: (1) the distinction between regulating neurotechnology and regulating its use, (2) the tension between developing situated regulations tailored to the local context and harmonizing local and international regulations and (3) the relationship between legal and ethical norms.

3 AI and Neurotechnology Regulation: Towards a Comparative Analysis

3.1 Deflection of Technology Liability

The deflection of technology liability highlights a pivotal aspect of legislative debates: the evolution from defining AI as a technology to questioning whether AI itself should be subject to regulation. While there was no universal consensus, a significant number of participants framed AI as a neutral tool—a mere object devoid of intentions. This perspective argued that responsibility should rest with human users rather than the technology itself, suggesting that existing legislation already governs human actions sufficiently.

The authors notice that in its initial draft, the Chilean Chamber of Deputies (CD) incorporated a definition of AI based on the European AI Act, conceptualizing it as a collection of computational techniques. However, this broad definition faced persistent criticism. Critics argued that it was so encompassing that even simple tools, like calculators, could fall under its scope. From a computer science perspective, this vagueness was problematic, conflating all computational systems with AI, thereby diluting the focus on tools exhibiting human-like intelligence. Key objections revolved around the portrayal of AI as functionally neutral, with its effects determined entirely by human agency. Analogies like "a hammer can build a house or attack a person" were used to argue that AI itself is neither inherently harmful nor beneficial—it depends on how individuals wield it. This perspective, prominent among industry representatives, emphasized regulating human misuse rather than the technology itself.

Conversely, legal scholars in Senate working groups advocated for broader, flexible definitions capable of adapting to the evolving nature of technology. Drawing from legal principles, they argued that rigid definitions risk excluding future phenomena that could require regulation. This perspective aligned with global trends, including Europe's adjustments to generative AI and anticipatory regulation for potential scenarios like general AI. The debates often revealed a tension between the technical precision demanded by computer scientists and the broader societal concerns raised by social scientists and legal experts. For example, efforts to highlight risks like misinformation and algorithmic bias were frequently reframed into technical language, leading to a disconnect. Computer scientists often dismissed these concerns as impractical or overly simplistic, stalling meaningful discussions on AI's social impacts.

The debate between technology as neutral or value-laden has also been raised during the drafting of the neurotechnology bill. In this case, there was a consensus between scholars and policymakers involved in the debate that the bill's aim should not be to regulate the technologies themselves, but their research, medical and commercial applications (Amunátegui Perelló, 2022). However, a key aspect of the bill may suggest otherwise. From an application-focused approach, one would expect to find distinct sets of rules oriented to the particularities of each domain. Nevertheless, a key article seems to establish regulations for medical technologies as the standard for the regulation of *all uses* of neurotechnology.

Article 7 of Bill 13828-19 states that all neurotechnologies should be registered by the Instituto de Salud Pública (ISP, *Institute of Public Health*)—the Chilean equivalent to the US's Food and Drug Administration (FDA). This registration requires that independently of whether the technology is intended for medical use or not, its effectiveness and safety would be tested in the same way as medical technologies. Medical testing studies must be performed to provide evidence regarding potential adverse effects. Depending on this evidence, The ISP may assign a qualification to each registered neurotechnology "according to their impact on people's health" (Article 8). In this sense, the law takes a medical standpoint and demands that non-clinical devices comply with the rules of therapeutic instruments. Crucially, no other consumer-oriented kind of device currently requires such a standard in Chile (Fuentes et al., 2025).

The nature of the assessment of the ISP seems to be aligned with the application-focused approach. The developers are required to state the intended uses of the technology, and the bill declares certain uses as explicitly prohibited. This includes influencing human conduct or extracting data without the user's consent, exploiting weaknesses, or affecting the neuroplasticity of vulnerable groups, children, and young adults. Some of these prohibitions were inspired by the European Proposal of Harmonized Rules on Artificial Intelligence, while others, such as the one regarding neuroplasticity, refer to a specific problem of neurotechnologies (Amunátegui Perelló, 2022).

Despite this apparent focus on the device applications, treating all neurotechnologies as medical seems to contradict their intrinsic neutrality. ISP registration process is aimed at preventing specific risks that may be unique to technologies intended for medical uses. However, Article 7 seems to presuppose that in the case of neurotechnologies, these risks are independent of the intended use. Is this a reasonable assumption? It may be in the case of 'stimulation' or 'modulation' neurotechnologies, technologies that influence neural activity through electrical or magnetical stimulation. Neural processes underlying psychological function are highly complex and, in most cases, still poorly understood. Therefore, potential harms to mental health may be much more difficult to predict than in the case of devices that influence the mind through the usual sensory pathways (e.g., digital technologies). Additionally, the physical interaction between the device and the brain exposes users to potential physical harm, a risk that is absent in other mind-related technologies. These risks may ground the idea that a high degree of protection is due to technology users independently of the intended use of the device.

However, this assumption may not be reasonable in the case of 'mind-reading' and non-invasive or non-implanted neurotechnologies. For instance, consider EEG devices. While being in contact with the scalp, EEG electrodes pick up neural signals by interacting with the waves of ions produced by neural activity (Yeh, 2012). Prima facie, all the physical and mental health risks mentioned above are not present in this kind of device-brain interaction. Furthermore, this is one of the technologies most widely used for non-medical uses, including education and labor-related applications (Muhl & Andorno, 2023; Privitera & Du, 2022). In this case, the reason to regard these technologies as intrinsically medical could be related to the specially sensitive nature of the neural data they collect (Wajnerman-Paz, 2022). Perhaps all neural data can be classified as medical data because of the medical information it can be inferred from it. This would justify treating all the 'reading' neurotechnologies as medical regardless of intended applications.

We agree that this is a plausible line of reasoning. However, it implies that many other technologies, including those targeted by the AI bill, should also be classified (and regulated) as medical (Wajnerman-Paz & López-Silva 2022). Behavioral data generated in digital platforms is processed by increasingly complex and powerful algorithmic techniques that have been proven to be successful in obtaining medical information. For instance, digital footprints of social media users may be used to identify a wide variety of medical conditions, such as postpartum depression disorder, PTSD, Anxiety, OCD, bipolar disorder, eating disorder, attention deficit, hyperactivity disorder, anxiety, and schizophrenia (Wongkoblap et al., 2017).

3.2 The Instrumentalization of Technology Policy

The instrumentalization of technology policy is the second source of controversy noted by the authors and it represents a core source of controversy in AI regulation, rooted in competing geopolitical and economic aspirations. In the Chilean legislative debates, stakeholders often framed regulatory approaches in terms of alignment with global paradigms, such as those established by the European Union (EU) or North America. Proponents of rapid regulation emphasized the symbolic importance of Chile being the first Latin American country to enact comprehensive AI laws, asserting this would position the nation as a regional leader in technology policy. For example, a legislator highlighted the need for Chile to navigate between regulatory models to align with global expectations, framing this ambition as indispensable given the rising ubiquity of AI across sectors.

Conversely, critics cautioned against the regulatory pressure stemming from such aspirations, arguing that premature legislation could inadvertently suppress innovation. One tech industry representative described the situation as a "regulatory burden" noting that strict adherence to international frameworks, such as the EU's AI Act, could impose excessive compliance costs, particularly for local enterprises. This view reflects broader concerns about the region's economic role as a consumer rather than a producer of AI technologies, with some stakeholders emphasizing the necessity of adopting pragmatic policies that foster infrastructural and developmental growth. The resulting tension highlights the dichotomy between the aspirations for "technological sovereignty"—as described by advocates for local innovation—and the perceived need to conform to external regulatory models that may not align with

Chile's socio-economic realities. These competing perspectives exemplify how political agendas and global positioning efforts shape and often constrain the discourse on AI governance.

The worry about overly restrictive regulation and the need to adapt to international regulation to avoid a negative impact on local marketing opportunities has also been discussed regarding neurotechnology (Rodríguez et al., 2021; Wajnerman-Paz, 2021). However, in this case, the international context is significantly different because a neurotechnology equivalent of AI regulations in the EU or the US did not exist when the discussion in Chile started. On the contrary, the Chilean bill was a pioneer attempt to implement a neurotechnology-centered framework that was intended to trigger a snowball effect in the rest of the world (Ruiz et al., 2024). Indeed, this seems to be what happened, as several Latin-American countries, including Mexico, Brazil, Argentina, and Colombia Colombia followed Chile in developing their regulatory proposals, often aligned with neurorights frameworks (Borbón & Ramírez Gómez, 2024). Prima facie, this context seems to provide a solution to the tension between technological sovereignty and the need to conform to external regulations, as Latin American countries would be protagonists in the development of international frameworks.

However, it is unlikely that this solution to the problem will work in the long term. The reason is that the neurorights frameworks on which most Latin American proposals are based have received significant criticism from academia (e.g. Borbón & Borbón, 2021; Bublitz, 2022; Hertz, 2023; Ligthart et al., 2023; Susser & Cabrera, 2023; Zuñiga-Fajuri et al., 2021). Symptomatic of these voices being heard are the more conservative proposals developed by leading organizations. For instance, current recommendations from the United Nations Human Rights Council (2024), OHCHR; OECD (2019), Council of Europe (2023), and the Inter-American Juridical Committee—OAS (2023), do not advocate for new rights, but rather make soft law and principle-based recommendations (Borbón, 2025).

This forces Latin America to face again the risk of losing technological sovereignty to fit into international regulation. Nevertheless, unlike what can be seen in the AI debate, the motivation to drop or reformulate their proposals could be grounded on much more than pragmatic or instrumental reasons. Criticism against some formulations of neurorights proposals is very widespread and well-founded and, to the best of our knowledge, has not received an appropriate response from their main advocates.

3.3 *The Moralization of Technology Use*

Lastly, the third source of controversy is called the moralization of technology use, which centers on the substitution of ethical frameworks for enforceable regulatory measures in AI governance. In Chile's legislative discourse, corporate actors and tech sector representatives frequently invoked ethical principles as an alternative to formal regulation, arguing that adherence to voluntary ethical guidelines could ensure socially responsible AI without imposing burdensome legal restrictions. This stance was exemplified by industry claims that companies already follow best practices in areas such as transparency, privacy, and non-discrimination, rendering additional legislative oversight unnecessary. For instance, representatives from multinational corporations like Google and Microsoft proposed self-regulation mechanisms as sufficient safeguards, emphasizing that ethical standards would promote innovation without stifling entrepreneurship.

Critics, however, identified this reliance on ethical discourse as a form of "ethics washing", where corporations use the language of ethics to resist legally binding accountability. Social scientists participating in the debates expressed concerns that deferring regulation in favor of ethical principles risks privileging industry interests over the protection of fundamental rights. One academic described this dynamic as a "remedy worse than the disease", suggesting that self-regulation fails to address systemic risks such as algorithmic bias or data misuse. The co-occurrence network in the discussions illustrated the separation between "ethics" and "regulation", signaling that ethics was often framed as a substitute rather than a complement to lawmaking. Ultimately, this controversy underscores how the invocation of ethics can serve as a strategic tool for delaying or diluting meaningful regulatory efforts, raising critical questions about the efficacy and accountability of self-regulation in the rapidly evolving AI landscape.

Unlike what can be seen in the AI debate, the voice of neurotechnology developers and companies has not been integrated into the local neurotechnology debate. Furthermore, neurotechnology companies have appeared in local media as adversaries rather than collaborators in the regulation and application of ethical standards to neurotechnologies (see Perello 2023; Muñoz et al., 2024). Moreover, the main neurorights frameworks developed in the US and the EU have had initially a strong focus on the legal protection of neurotechnology users (Ienca, 2021).

In the case of the neurotechnology debate, it seems that an opposing trend can be observed. Instead of minimizing the legal protection of users by relying more on ethical guidelines for developers and companies, it seems that the legal protection has been extended to domains that may belong to purely moral considerations. Just to mention a paradigmatic example, one of the most discussed neurorights was the right to equal access to neurotechnologies for mental enhancement or augmentation. As Borbón et al. (2021) argue, "this new right could pose the problem of a new burden for the State, for which it would be forced to finance, with public funds, the introduction of all the neurotechnologies that are developed." Alternatively, if this becomes a duty for companies, their financial interests would be undermined and therefore they would lose any incentive for technological innovation. The problem, as we see it, is that guaranteeing access to technologies is morally desirable but may belong to the realm of the supererogatory rather than to the mandatory, and therefore is not a norm that should be transformed into legislation.

Similarly, the neuroright to 'free will' grounds the mentioned prohibitions to technologically influencing behavior. Besides neurotechnology, digital technologies used for 'algorithmic nudging' have also been the focus of this concern. However, many scholars have noticed that it is very difficult to draw a clear line between technological persuasion and coercion, a line that can be turned into precise policy (Zohny et al.,

2023). Although it is morally desirable that people do not use technology to bypass the decision-making capacity of others to serve their purposes, it may be excessive to prohibit all forms of influence, many of which are a constitutive part of most of our daily social interactions.

4 Concluding Remarks

The regulatory challenges surrounding AI and neurotechnology in Chile reflect broader global tensions between innovation, ethical accountability, and legal oversight. While AI governance has sparked contentious debates on liability, policy instrumentalization, and ethical discourse, neurotechnology regulation presents distinct yet overlapping concerns, particularly regarding the classification of these technologies and their potential risks. This comparative analysis highlights how Chile's approach to regulating these emerging fields has been shaped by both international influences and local political dynamics, revealing key areas where regulatory frameworks must evolve to address their unique complexities.

A central issue in both AI and neurotechnology regulation is the deflection of technology liability. In the AI debate, the tendency to frame AI as a neutral tool has led to arguments that human users, rather than the technology itself, should bear responsibility for its harm. This perspective, largely driven by industry stakeholders, risks overlooking the systemic biases and emergent behaviors inherent in AI systems. Conversely, in the case of neurotechnology, the Chilean legislative framework appears to take a more interventionist stance by mandating the classification of all neurotechnologies under medical regulations, regardless of their intended use. This assumption contradicts the intrinsic neutrality of neurotechnologies and raises questions about whether the risks associated with specific applications warrant blanket regulatory measures that may hinder innovation and non-medical applications.

The instrumentalization of technology policy also plays a crucial role in shaping AI and neurotechnology regulation. In AI governance, debates have been influenced by the desire to align with global regulatory trends, particularly those set by the European Union, while also ensuring that Chile does not impose undue restrictions that could stifle local innovation. In contrast, Chile has positioned itself as a pioneer in neurotechnology regulation, with its neurorights framework inspiring legislative efforts across Latin America. However, this leadership role comes with its own challenges, as emerging criticisms of neurorights frameworks may force Chile to reconsider its regulatory approach. The evolving international stance—shifting from rights-based frameworks to softer regulatory guidelines—suggests that Chilean policymakers will need to navigate the tension between maintaining technological sovereignty and ensuring compatibility with global governance standards.

A further distinction between AI and neurotechnology regulation lies in the moralization of technology use. In AI governance, ethical principles are often invoked as an alternative to formal regulation, a trend that critics argue serves to delay meaningful oversight while allowing corporate interests to dominate the discourse. In contrast, the neurotechnology debate appears to exhibit the opposite tendency, where legal protections have been extended into areas that may be better suited to moral deliberation rather than strict legal enforcement. By critically assessing the interplay between local regulatory debates and international developments, this analysis highlights the importance of avoiding both regulatory overreach and laissez-faire approaches. As Chile continues to shape its AI and neurotechnology policies, it must navigate the fine line between fostering technological innovation and safeguarding fundamental rights. The insights gained from this comparative perspective offer valuable lessons for the broader global discourse on emerging technology governance, emphasizing the need for regulatory models that are flexible, equitable, and attuned to the sociotechnical realities of AI and neurotechnology across different social contexts.

References

Amunátegui Perelló, C. (2022). Natural and artificial neural networks: The Chilean legal framework. *Journal of Civil Law Studies, 14*(1), 11.

Borbón, D., & Borbón, L. (2021). A critical perspective on neurorights: Comments regarding ethics and law. *Frontier Human Neuroscience, 15*, 703121. https://doi.org/10.3389/fnhum.2021.703121

Borbón, D., & Ramírez-Gómez, J. A. (2024). Between politics and scholarship: The (un)settled debate over neurorights. *Frontiers in Political Science, 6*.https://doi.org/10.3389/fpos.2024.1335561

Borbón, D. (2025). Neuroética y neuroderechos ante las propuestas de neurointervenciones coercitivas con fines de mejora moral en ámbitos penales. *Cuadernos Electrónicos De Filosofía Del Derecho, 53*.https://doi.org/10.7203/CEFD.53.30241

Bublitz, J. C. (2022). Novel neurorights: From nonsense to substance. *Neuroethics, 15*, 7. https://doi.org/10.1007/s12152-022-09481-3

de Bruin, R. W. (2024). Co-regulation and AI-innovation: Principles for a sustainable framework fostering innovation and acceptance of AI. In M. I. Aldinhas Ferreira (Ed.), Producing artificial intelligent systems: The roles of benchmarking, standardisation and certification (pp. 119–140). Springer Nature. https://doi.org/10.1007/978-3-031-55817-7_8

Finocchiaro, G. (2024). The regulation of artificial intelligence. *AI & Society, 39*, 1961–1968. https://doi.org/10.1007/s00146-023-01650-z

Fuentes, J., Ramos Vergara, P., Wajnerman-Paz, A. López-Silva, P., & Álamos, F., (2025) A situated approach to neurorights legislation in Chile. In, López-Silva, P. (Ed.), *Contextualizing neuroprotection: Latin American perspectives on the impact of neurotechnological development in life and society*, Springer.

Hertz, N. (2023). Neurorights—Do we need new human rights? A reconsideration of the right to freedom of thought. *Neuroethics,16*, 5. https://doi.org/10.1007/s12152-022-09511-0

Humeres, M., Cotoras, D., Moretti, R., Oyarzún-Merino, I., Correa, T., & López, C. (2025). Cooling down AI regulation controversies: Three closure processes in the Chilean legislative arena. *Big Data Society, 12*(1). https://doi.org/10.1177/20539517241311067

Ienca, M. (2021). On neurorights. *Front Human Neuroscience, 15*. https://doi.org/10.3389/fnhum.2021.701258

Judge, B., Nitzberg, M., & Russell, S. (2024). When code isn't law: Rethinking regulation for artificial intelligence, *Policy and Society*. https://doi.org/10.1093/polsoc/puae020

Kane, A. (2021). Regulating AI: Considerations that apply across domains. In J. von Braun, M. S. Archer, G. M. Reichberg, & M. Sánchez Sorondo (Eds.), Robotics, AI, and humanity: Science, ethics, and policy (pp. 251–259). Springer International Publishing. https://doi.org/10.1007/978-3-030-54173-6_21

López-Silva, P. & Madrid, R. (2021). Sobre la Conveniencia de incluir los Neuroderechos en la Constitución o en la Ley. *Revista Chilena de Derecho y Tecnología, 10*(1), 49–72. https://doi.org/10.5354/0719-2584.2021.56317

Laptev, V. A., & Feyzrakhmanova, D. R. (2024). Application of artificial intelligence in justice: Current trends and future prospects. *Human-Centric Intelligent Systems, 4*(3), 394–405. https://doi.org/10.1007/s44230-024-00074-2

Ligthart, S., Ienca, M., Meynen, G., Molnar-Gabor, F., Andorno, R., Bublitz, C., Catley, P., Claydon, L., Douglas, T., Farahany, N., Fins, J., Goering, S., Haselager, P., Jotterand, F., Lavazza, A., McCay, A., Wajnerman Paz, A., Rainey, S., Ryberg, J., & Kellmeyer, P. (2023). Minding rights: Mapping ethical and legal foundations of 'neurorights.' *Cambridge Quarterly of Healthcare Ethics, 2023*, 1–21.

López-Silva, P. & Madrid, R. (2022). Protegiendo la Mente: Un análisis al Concepto de lo Mental en la Ley de Neuroderechos. *Revista de Humanidades de Valparaíso, 20*, 101–117. https://doi.org/10.22370/rhv2022iss20pp101-117

López-Silva, P. & Madrid, R. (2024). ¿Es la noción de 'neuroderecho' un concepto jurídico indeterminado? *Lus et Praxis, 30*(3). https://doi.org/10.4067/S0718-00122024000300004

Muhl, E., & Andorno, R. (2023). Neurosurveillance in the workplace: Do employers have the right to monitor employees' minds? *Frontiers in Human Dynamics, 5*, 1245619.

Muñoz, J. M., Marinaro, J. Á., Iglesias, J. A., Sánchez, M., Monti, N., Colombara, C., & Girardi, G. (2024). Effects of the first successful lawsuit against a consumer neurotechnology company for violating brain data privacy. *Nature Biotechnology*, 1–2.

Palaniappan, K., Lin, E. Y. T., & Vogel, S. (2024). Global regulatory frameworks for the use of artificial intelligence (AI) in the healthcare services sector. *Healthcare, 12*(5), 562. https://doi.org/10.3390/healthcare12050562

Perello, C. A. (2023). Neurorights.: The World's first court case. *Intellectual Property Technology Law Journal, 28*, 123.

Privitera, A. J., & Du, H. (2022). Educational neurotechnology: Where do we go from here? *Trends in Neuroscience and Education, 29*, Article 100195.

Rodríguez, D. A. B., Rodríguez, L. F. B., & Bustamante, M. A. L. (2021). NeuroRight to equal access to mental augmentation: Analysis from posthumanism, law and bioethics. *Revista Iberoamericana De Bioética, 16*, 01–15.

Ruiz, S., Valera, L., Ramos, P., & Sitaram, R. (2024). Neurorights in the constitution: From neurotechnology to ethics and politics. *Philosophical Transactions B, 379*(1915), 20230098.

Staszczyk, P. (2024). Chapter 3: Navigating the AI landscape in the EU: fostering innovation while upholding ethical principles. https://www.elgaronline.com/edcollchap-oa/book/9781035337934/book-part-9781035337934-10.xml

Susser, D., & L.Y. Cabrera. (2023). Brain data in context: Are new rights the way to mental and brain privacy?. *AJOB Neuroscience*, 1–12.

Tschider, C., Compagnucci, M. C., & Minssen, T. (2024). The new EU–US data protection framework's implications for healthcare. *Journal of Law and the Biosciences, 11*(2), lsae022. https://doi.org/10.1093/jlb/lsae022

Wajnerman, A., & López-Silva, P. (2022). Mental privacy and neuroprotection: An open debate. *Protecting the mind: Challenges in law, neuroprotection, and neurorights* (pp. 141–155). Springer International Publishing.

Wajnerman-Paz, A. (2021). We need to regulate mind-reading tech before it exists. Rest of World. https://restofworld.org/2021/chile-neuro-rights/

Wongkoblap, A., Vadillo, M. A., & Curcin, V. (2017). Researching mental health disorders in the era of social media: systematic review. *Journal of Medical Internet Research, 19*(6), e228. https://doi.org/10.2196/jmir.7215

Yeh, M. (2012). Digital EEG. In T. Yamada & E. Meng (Eds.), *Practical guide for clinical neurophysiologic testing: EEG* (pp. 54–62). Lippincott Williams & Wilkins.

Zaidan, E., & Ibrahim, I. A. (2024). AI governance in a complex and rapidly changing regulatory landscape: A global perspective. *Humanities Social Science Communications, 11*, 1121. https://doi.org/10.1057/s41599-024-03560-x

Zohny, H., Lyreskog, D. M., & Singh I. (2023). The mystery of mental integrity: Clarifying its relevance to neurotechnologies. *Neuroethics, 16*, 20. https://doi.org/10.1007/s12152-023-09525-2

Zuñiga-Fajuri, A., Villavicencio, L., & Zaror-Millares, D. (2021). Neurorights in Chile: Between neuroscience and legal science. In M. Hevia (Ed.), *Regulating neuroscience: Transnational legal challenges* (pp. 165–179). Elsevier.

Neurotechnological Progress and the Quest for Contextualized Regulations

Pablo López-Silva

Abstract This chapter offers a brief commentary on the main contributions of Latin American researchers to debates about the ethical, socio-economic, and geopolitical impact of the use of neurotechnologies. Latin America's vulnerability to limited access and development of neurotechnologies underscores the risk of exacerbating existing inequalities. Yet, this vulnerability also provides valuable insights into the consequences of rushed regulation, the importance of interdisciplinary collaboration, and the need for technological education to address socio-economic and epistemic disparities.

Keywords Neuroprotection · Neurorights · Latin America · Neurotechnology

In his *Collected Essays*, Aldous Huxley (1959) wrote, "*That men do not learn very much from the lessons of history is the most important of all the lessons of history*" (p. 308). This aphorism feels especially relevant when examining the impact of neurotechnological development on both local and global aspects of our lives. Huxley's pessimism—inherited from the original source of the idea, Georg Hegel—becomes particularly striking as we witness history repeating itself in many global affairs in recent times. The same patterns emerge in history again and again. The case of neurotechnology is particularly interesting in this regard, as it endows us with methods and techniques to intervene in the brain on an unprecedented scale.

P. López-Silva (✉)
School of Psychology, Universidad de Valparaíso, Valparaiso, Chile
e-mail: pablo.lopez.silva@gmail.com

Millennium Institute for Research in Depression and Personality (MIDAP-Chile), Santiago, Chile

Universidad de Valparaíso Research Centre in Cognition and Language Development, Valparaíso, Chile

Valparaíso Institute for Complex Systems, Valparaíso, Chile

Collaborating Centre for Value-Based Research, St Catherine's College, University of Oxford, Oxford, UK

P. López-Silva (ed.), *Contextualizing Neuroprotection: Latin American Perspectives on the Impact of Neurotechnological Development in Life and Society*, Ethics of Science and Technology Assessment 54, https://doi.org/10.1007/978-3-031-96055-0_16

Thus, the introduction of this type of technology into society opens up a number of possibilities. For a variety of conceptual, legal, and practical reasons, the examination of how societies should deal with neurotechnological development has not been without challenges. One might assume that the values humans seek to protect are similar across cultures, but this is not entirely the case. Similarly, different societies strive to protect their, seemingly equal, fundamental values in diverse—and sometimes opposing—ways.

The irruption of the neurorights crusade in Latin America has made clear how some ideas and actions have been introduced in the region without proper consideration of the context. Regardless of the actual outcomes of this ongoing process, a future where the privacy of our inner lives is open to intrusion and control is no longer a dystopian fantasy; it has become a real ethical horizon in our current world. Becoming aware of this threat has been one of the many issues driving the debate in the region. Nobody could deny that neurotechnologies have enabled the development of powerful ways to treat serious diseases, improving the well-being of many around the world. However, at the same time, this progress has led to a number of novel and ethically relevant scenarios and, with this, to a number of far-from-resolved discussions.

In this context, let me go back to what Serrano (in *this volume*) asks when examining the debate about neurorights and neuroprotection in Latin America. Due to socio-economic, socio-historical, and socio-political reasons, it is not controversial to say that Latin American countries are not where globally relevant neurotechnologies are systematically developed (see Yuste et al., 2021). However, quite paradoxically, Chile is the first country in the world to approve a constitutional amendment explicitly protecting neural data in light of the increasing massification of neurotechnologies. This raises the question, as Serrano does, of what meaningful contributions countries not directly involved in the research and development of cutting-edge neurotechnology can offer to the ongoing debate. Often, neurotechnological innovations are developed by wealthier nations (or "Global North countries"). Consequently, one might argue that neurotechnological regulation and its associated arguments originating in Latin America could overlook core aspects of the issue, as they may be disconnected from the actual processes of neurotechnology creation and implementation. This could be partly true. However, it might not necessarily be a bad thing. Latin American countries can offer a unique and valuable perspective on certain aspects of the current debate surrounding neurorights and neuroprotection. For example, consider the lack of economic and political power that neurotech companies hold in these countries compared to wealthier economies. Rather than protecting investments or becoming entangled in complex socio-political and economic dynamics, the focus of the discussion in Latin America has been on the potential negative consequences of neurotechnology—particularly on those that, should they arise, these countries would be ill-equipped to mitigate due to the socio-economic and epistemic inequalities existing in the region. While it is essential to avoid exaggerating these concerns, it is equally crucial not to minimize the potential for adverse effects, especially considering that Latin America is home to some of the most unequal societies in the world.

In this context, the insights from Latin American researchers regarding the ethical implications of neurotechnological development and commercialization could contribute significantly to identifying and addressing overlooked or under-represented issues in the international debate. The contributions to this compilation are proof of this. All the issues treated in this compilation could especially inform debates occurring in societies with unequal distributions of economic and epistemic resources—in other words, most countries in the world. For this reason, the peripheral role of Latin American countries in the process of neurotechnological development and commercialization can be considered generative in many respects. In Chile, for example, it has prompted consideration of the negative consequences that rushed regulations on neurotechnological use in medicine could have on industry and, consequently, on a country's economic and research development (Ruiz et al., 2021). In the broader region, it serves to show the world how certain things could be done and to warn against how not to do others.

Peripheral countries, such as those in Latin America, face the realistic threat of limited or no access to neurotechnology, while simultaneously lacking the capacity to develop accessible and competitive neurotechnologies that could address the disparity between the region and developed countries. This vulnerability to global changes offers a nuanced and unique perspective, as the lack of development and access to neurotechnologies could exacerbate existing socio-economic, epistemic, and geopolitical disadvantages. Even if it is true that Latin American countries currently play a peripheral role in developing cutting-edge neurotechnologies with global impact, the perspectives emerging from these countries are capable of offering valuable insights gained from the direct experience including missteps and the exploration of uncharted avenues.

As discussed in Chap. 1, being a pioneer in a field presents both advantages and challenges. While it allows for innovation and the establishment of novel paradigms and ideas, it also entails navigating uncharted territory, facing resistance, and lacking established conceptual frameworks for support. As the old saying goes, "*Experience is the mother of wisdom*". *Contextualizing Neuroprotection: Latin American Perspectives on the Impact of Neurotechnological Development on Life and Society* has been an attempt to conceptualize and systematize what Latin American researchers have learned from their experiences—both in terms of what to do and what not to do—when it comes to debates about regulating and thinking about neurotechnological development. It is a call to recognize the importance of interdisciplinary collaboration without epistemic trespassing, to acknowledge how the political instrumentalization of relevant debates can delay necessary societal changes, and to emphasize the need for technological education in societies, so as to address epistemic inequalities that could harm citizens' lives. I sincerely hope that our work can offer insightful ideas to motivate collaboration in a globalized world, where the actions of some can affect us all.

References

Huxley, A. (1959). *Collected essays*. Harper & Brothers.

Ruiz, S., Ramos-Vergara, P., Rodrigo, C., Altermatt, F., Von-Bernhardi, R., Mauricio, C., et al. (2021). Negative effects of the patients' rights law and neuro-rights bill in Chile. *Revista Médica De Chile, 149*, 439–446. https://doi.org/10.4067/s0034-98872021000300439

Yuste, R., Genser, J., & Herrmann, S. (2021). It's time for neuro-rights: New human rights for the age of neurotechnology. *Horizons: Journal of International Relations and Sustainable Development, 18*, 154–165. https://www.jstor.org/stable/48614119